Lecture Notes in Computer Science   16126

Founding Editors

Gerhard Goos
Juris Hartmanis

Editorial Board Members

Elisa Bertino, *Purdue University, West Lafayette, IN, USA*
Wen Gao, *Peking University, Beijing, China*
Bernhard Steffen , *TU Dortmund University, Dortmund, Germany*
Moti Yung , *Columbia University, New York, NY, USA*

The series Lecture Notes in Computer Science (LNCS), including its subseries Lecture Notes in Artificial Intelligence (LNAI) and Lecture Notes in Bioinformatics (LNBI), has established itself as a medium for the publication of new developments in computer science and information technology research, teaching, and education.

LNCS enjoys close cooperation with the computer science R & D community, the series counts many renowned academics among its volume editors and paper authors, and collaborates with prestigious societies. Its mission is to serve this international community by providing an invaluable service, mainly focused on the publication of conference and workshop proceedings and postproceedings. LNCS commenced publication in 1973.

Marius Erdt · Yufei Chen · Stefan Wesarg ·
Klaus Drechsler · Moti Freiman ·
Sarina Thomas · Samah Khawaled
Editors

Clinical Image-Based Procedures

14th International Workshop, CLIP 2025
Held in Conjunction with MICCAI 2025
Daejeon, South Korea, September 23, 2025
Proceedings

Editors
Marius Erdt
Independent Researcher
Singapore, Singapore

Stefan Wesarg
Fraunhofer Institute for Computer Graphics
Research (IGD)
Darmstadt, Germany

Moti Freiman
Technion – Israel Institute of Technology
Haifa, Israel

Samah Khawaled
Weill Cornell Medicine and Cornell Tech
New York, NY, USA

Yufei Chen
Tongji University
Shanghai, China

Klaus Drechsler
Aachen University of Applied Sciences
Aachen, Germany

Sarina Thomas
University of Oslo
Oslo, Norway

ISSN 0302-9743　　　　　　　ISSN 1611-3349　(electronic)
Lecture Notes in Computer Science
ISBN 978-3-032-05478-4　　　ISBN 978-3-032-05479-1　(eBook)
https://doi.org/10.1007/978-3-032-05479-1

© The Editor(s) (if applicable) and The Author(s), under exclusive license
to Springer Nature Switzerland AG 2026
Chapter "Ablate them all: A trajectory planning for concurrent percutaneous ablation of multiple tumors" is licensed under the terms of the Creative Commons Attribution 4.0 International License (http://creativecommons.org/licenses/by/4.0/). For further details see license information in the chapter.

This work is subject to copyright. All rights are solely and exclusively licensed by the Publisher, whether the whole or part of the material is concerned, specifically the rights of translation, reprinting, reuse of illustrations, recitation, broadcasting, reproduction on microfilms or in any other physical way, and transmission or information storage and retrieval, electronic adaptation, computer software, or by similar or dissimilar methodology now known or hereafter developed.
The use of general descriptive names, registered names, trademarks, service marks, etc. in this publication does not imply, even in the absence of a specific statement, that such names are exempt from the relevant protective laws and regulations and therefore free for general use.
The publisher, the authors and the editors are safe to assume that the advice and information in this book are believed to be true and accurate at the date of publication. Neither the publisher nor the authors or the editors give a warranty, expressed or implied, with respect to the material contained herein or for any errors or omissions that may have been made. The publisher remains neutral with regard to jurisdictional claims in published maps and institutional affiliations.

This Springer imprint is published by the registered company Springer Nature Switzerland AG
The registered company address is: Gewerbestrasse 11, 6330 Cham, Switzerland

If disposing of this product, please recycle the paper.

Preface

The 14th International Workshop on Clinical Image-based Procedures: Towards Holistic Patient Models for Personalized Healthcare (CLIP 2025) was held in Daejeon, South Korea, on September 23rd, 2025. CLIP 2025 was organized in conjunction with the International Conference on Medical Image Computing and Computer Assisted Intervention (MICCAI).

The focus of our workshop is on translational research and it provides a forum for scientific work applied to clinical practice. Furthermore, holistic personalised patient models become increasingly important for healthcare. Such personalised models combine medical image data from multiple modalities with other patient data, such as omics, demographics, and electronic health records for better diagnosis and treatment. Since 2019, CLIP has placed special emphasis on this area of research.

CLIP 2025 received 17 submissions. All submitted papers were double-blindly peer-reviewed by at least two experts, and 11 papers were finally accepted for presentation at the workshop and included in this book.

We are very grateful to MICCAI for their long-standing support in providing the platform for our workshop. We also express our deepest gratitude to all the authors and our program committee members who made CLIP 2025 a success.

August 2025

Marius Erdt
Yufei Chen
Stefan Wesarg
Klaus Drechsler
Moti Freiman
Sarina Thomas
Samah Khawaled

Organization

Organizing Committee

Erdt, Marius	Independent Researcher, Singapore
Chen, Yufei	Tongji University, China
Wesarg, Stefan	Fraunhofer Institute for Computer Graphics Research (IGD), Germany
Drechsler, Klaus	Aachen University of Applied Sciences, Germany
Freiman, Moti	Technion – Israel Institute of Technology, Israel
Thomas, Sarina	University of Oslo, Norway
Khawaled, Samah	Weill Cornell Medicine and Cornell Tech, USA

Reviewers

Egger, Jan
Hoßbach, Martin
Luijten, Gijs
Karpate, Yogesh
Kausch, Lisa
Nizam, Nusrat
Park, Seungbin
Rheude, Tillmann
Tiago, Cristiana
Wan, Cheng
Wetzer, Elisabeth
Wickstrøm, Kristoffer
Zidowitz, Stephan

Contents

2D/3D Registration of Acetabular Hip Implants Under Perspective
Projection and Fully Differentiable Ellipse Fitting 1
  Yehyun Suh, J. Ryan Martin, and Daniel Moyer

ECG Report Generation with Diagnostic Knowledge Enhanced Prompt
Learning 11
  *Panpan Fan, Xiaodong Yue, Yufei Chen, Zhipeng Wei, Jie Shi,
  and Zhikang Xu*

TileDVP: Decoding the Tissue Proteome from H&E Images 21
  *Émilie Mathian, Lukas Oldenburg, Eduard Chelebian, Lisa Schweizer,
  Gijs Zonderland, Kristian Egebjerg, Ishani Ummat, Andreas Mund,
  and Maximillian T. Strauss*

Domain-Specialized Interactive Segmentation Framework for Meningioma
Radiotherapy Planning 32
  Junhyeok Lee, Han Jang, and Kyu Sung Choi

AI-Driven Multimodal TMJ Patient Modeling: From Unstructured Notes
to Precision Treatment 42
  *Alban Gaydamour, Enzo Tulissi, Claudia Mattos, Rodrigo Teixeira,
  Maxwell Shin, Adam Hershey, Anabelle Kwon, Felicia Miranda,
  Marcela Gurgel, Selene Barone, Aron Aliaga, Marilia Yatabe,
  Paulo Zupelari, Marina Zupelari, David Hanauer, Nina Hsu,
  Steve Pieper, Eduardo Caleme, Jonas Bianchi, Joao Goncalves,
  Daniela Goncalves, Lawrence Wolford, Antonio Ruellas,
  Juan Prieto, Tengfei Li, Hongtu Zhu, Runpeng Dai, Martin Styner,
  Najla Al Turkestani, Alexandre F. DaSilva, and Lucia Cevidanes*

OrificeNet: Automatic Concealed Orifice Detection from Microscope
Imagery with CBCT-Guided Refinement 53
  Kefan Zhou, Yufei Chen, Wei Liu, Qiyun Shen, and Qi Zhang

Ablate Them All: A Trajectory Planning for Concurrent Percutaneous
Ablation of Multiple Tumors 65
  Adela Lukes, Stefano Fogarollo, Reto Bale, and Wolfgang Freysinger

NEURAL: Attention-Guided Pruning for Unified Multimodal
Resource-Constrained Clinical Evaluation 75
  Devvrat Joshi and Islem Rekik

Interpreting CT-Scans with CLIP: An Explorative Study of Attribution
Methods for 3D Vision-Language Models 87
   David Avedis Injarabian, Joonas Ariva, Hendrik Šuvalov,
   and Dmytro Fishman

Automated Constraint-Aware X-ray View Planning for Vascular
Interventions Using Preoperative CTA 97
   Baochang Zhang, Abdelkader Saad, Heribert Schunkert,
   and Nassir Navab

Author Index 107

2D/3D Registration of Acetabular Hip Implants Under Perspective Projection and Fully Differentiable Ellipse Fitting

Yehyun Suh[1,2,3], J. Ryan Martin[4], and Daniel Moyer[1,2,3]( )

[1] Department of Computer Science, Vanderbilt University, Nashville, TN 37235, USA
[2] Vanderbilt Institute of Surgery and Engineering, Nashville, TN 37235, USA
[3] Vanderbilt Lab for Immersive AI Translation, Nashville, TN 37235, USA
[4] Department of Orthopaedic Surgery, Vanderbilt University Medical Center, Nashville, TN 37232, USA
{yehyun.suh,daniel.moyer}@vanderbilt.edu

Abstract. This paper presents a novel method for estimating the orientation and the position of acetabular hip implants in total hip arthroplasty using full anterior-posterior hip fluoroscopy images. Our method accounts for distortions induced in the fluoroscope geometry, estimating acetabular component pose by creating a forward model of the perspective projection and implementing differentiable ellipse fitting for the similarity of our estimation from the ground truth. This approach enables precise estimation of the implant's rotation (anteversion, inclination) and the translation under the fluoroscope-induced deformation. Experimental results from both numerically simulated and digitally reconstructed radiograph environments demonstrate high accuracy with minimal computational demands, offering enhanced precision and applicability in clinical and surgical settings. Code: https://github.com/yehyunsuh/Acetabular-Cup-Pose-Estimator.

Keywords: 2D/3D Registration · Acetabular Hip Implant · Circular Object Pose Estimation · Fully Differentiable Ellipse Fitting

1 Introduction

Total Hip Arthroplasty (THA) is a surgical procedure that replaces damaged hip joint cartilage and bone with artificial components [2]. Complications and post-procedure negative outcomes have in part been ascribed to the orientation of the acetabular component ("hip joint cup" or "cup") relative to the patient's natural position [10].

It is therefore important to provide intra-operative tracking and pose-estimation for the cup orientation. As the cups have a known and relatively simple geometry (effectively hemispheres of known radius), and have high radio-density, fluoroscopy is the gold-standard option for localization and pose estimation. In the present work, we estimate the 3D pose of the implanted hemispheres given the observed ellipse in 2D fluoroscopy images.

This task would be analytically solvable using elementary mathematics for projections along standard Euclidean axes (orthographic projection). However, image intensifier fluoroscopes have a perspective projection geometry. Images are formed by projecting rays from the X-ray source (a point) to the collection panel (a plane), which results in deformations at coordinates away from the origin. Current methods are largely focused on flat-plate or film collection [9] cases, or stereotaxis (parallax) effects [12]. The former have significant error when applied to hemispheres projected with this deformation. The latter requires multiple image collections from varying points of view, and error in C-arm (fluoroscope mount) motion or control induces further error.

2D/3D registration offers solution for estimating cup pose by generating a moving image from digitally reconstructed radiographs (DRR) [6,14] and comparing it to a target image. While well-established, its practical application is limited by the computational inefficiency of repeatedly projecting the 3D implant model and the limited discriminative power of similarity calculations, which reduce registration accuracy. Additionally, requiring the exact 3D model-often unavailable in retrospective studies-further limits feasibility. However, this challenge of estimating a cup pose is a special case that can be addressed more efficiently. The method in this study overcomes these issues, enabling robust, real-world registration without exhaustive 3D modeling.

This paper proposes a method for estimating the acetabular hip implant pose in fluoroscopy images. Our contributions are:

- pose estimation based on a perspective projection (cone-beam projection), avoiding the error in orthographic projection methods
- elimination of the need to project a full 3D model of the implant, significantly reducing computational costs
- error calculation using geometric descriptors of ellipses instead of landmark positions, solving a correspondence symmetry problem.
- a fully differentiable ellipse fitting process, enabling gradient-based optimization for precise pose estimation.

We demonstrate the effectiveness of our method by comparing it with orthographic projection for orientation estimation, and intensity-based and embedding-based 2D/3D registration methods for pose estimation. We also show the robustness in read-world data by implementing our method on cadaver data.

2 Method

Our objective is to recover the pose of the cup from intra-operative fluoroscopy. The pose is defined by five variables: anteversion/extra planar rotation θ (rotation into/out of the image plane), inclination/rotation within the image plane

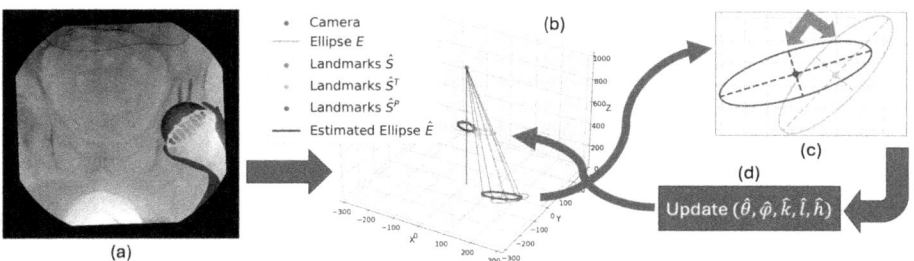

Fig. 1. Registration pipeline. (a) **Segment** the ellipse (orange cross-line), **extract** landmarks S^P (orange point) on the edge of the ellipse, and **calculate** ellipse E (orange ellipsoidal-line) by fitting an ellipse on S^P. (b) Rotate and translate landmarks $\hat{S}$ (red) to $\hat{S}^T$ (light blue), **project** to obtain $\hat{S}^P$ (purple), and **calculate** ellipse $\hat{E}$ (purple line) by fitting an ellipse on $\hat{S}^P$. (c) Calculate the difference (red arrow) between E and $\hat{E}$. (d) Update variable $(\hat{\theta}, \hat{\varphi}, \hat{k}, \hat{l}, \hat{h})$. Repeat process (b), (c), (d) until convergence. (Color figure online)

φ, in-plane translation (k, l), and extra-planar translation h of the cup. By convention these rotations are of the cup and not the imaging system/detector. The cups are assumed to be metal hemispheres with high radiodensity/image contrast, and annotated inner faces. Annotation of these structures in fluoroscopy can be performed automatically using e.g. a neural network [11,13], which we demonstrate in several experiments but do not describe for brevity.

Our method is composed of the following steps (shown graphically in Fig. 1):

1. Observe an ellipse E using landmarks (denoted S^P) observed from the fluoroscopy by fitting an ellipse to it.
2. Given a 3D pose of landmarks ($\hat{S}$), estimate the projected landmarks ($\hat{S}^P$) in the image plane and estimate the nominal projected ellipse ($\hat{E}$).
3. Calculate error between the observed ellipse and the nominal projected ellipse parameters, then update the 3D pose via gradient descent (or another optimization method).
4. Repeat from Step 2 until convergence.

Notably, this method does not require correspondence between landmarks in S^P and $\hat{S}^P$ in the image plane or in the estimated 3D pose landmarks $\hat{S}$.

2.1 Ellipse Fitting from Landmarks

Fitting the ellipse E and $\hat{E}$ from landmarks S^P and $\hat{S}^P$ (Fig. 1(a)) uses methods from Fitzgibbon et al. [4] to form the least squares ellipse fit. This involves transforming the Euclidean coordinates of the landmarks into a six-dimensional space that parameterizes the implicit form of the ellipse equation:

$$Ax^2 + Bxy + Cy^2 + Dx + Ey + F = 0. \tag{1}$$

Each landmark coordinates (x_i, y_i) in S^P is transformed to a new vector:

$$[x_i, y_i] \mapsto [x_i^2, x_i y_i, y_i^2, x_i, y_i, 1]. \tag{2}$$

We concatenate these new vectors as rows in a data matrix X_{ep}. From this, the scatter matrix M is calculated as:

$$M = X_{ep}^T X_{ep}. \tag{3}$$

The generalized eigenvalue problem is solved using singular value decomposition (SVD) on the inverse of the scatter matrix, M^{-1}:

$$U, S, V = \text{SVD}(M^{-1}). \tag{4}$$

The first column of U are the coefficients of the ellipse: $A, B, C, D, E, F = U_{:,1}$. While the implicit parametrization uniquely describes each ellipse, error calculated in these parameters does not nicely correspond with intuitive notions of geometric error (e.g., Hausdorff differences between curves in the image plane). Thus, we convert coefficients to the standard parameterization. The center (x, y) of the ellipse is:

$$x = \frac{C \cdot D - \frac{B}{2} \cdot E}{2\left(\left(\frac{B}{2}\right)^2 - A \cdot C\right)}, y = \frac{A \cdot E - \frac{B}{2} \cdot D}{2\left(\left(\frac{B}{2}\right)^2 - A \cdot C\right)}. \tag{5}$$

For ease of notation, we can define a series of auxiliary variables:

$$\mu = \frac{1}{A \cdot x^2 + 2 \cdot \frac{B}{2} \cdot x \cdot y + C \cdot y^2 - F}, \tag{6}$$

and $m_{11} = \mu \cdot A$, $m_{12} = \mu \cdot \frac{B}{2}$, and $m_{22} = \mu \cdot C$. The semi-major and minor axes a and b are expressed in terms of these variables:

$$a = \frac{1}{\frac{1}{2}\left(m_{11} + m_{22} + \sqrt{(m_{11} - m_{22})^2 + 4 \cdot m_{12}^2}\right)} \tag{7}$$

$$b = \frac{1}{\frac{1}{2}\left(m_{11} + m_{22} - \sqrt{(m_{11} - m_{22})^2 + 4 \cdot m_{12}^2}\right)}. \tag{8}$$

The orientation of the ellipse is calculated as:

$$\alpha = \frac{1}{2} \cdot \text{atan2}\left(-2 \cdot \frac{B}{2}, C - A\right) \times \frac{180}{\pi}, \tag{9}$$

and the parameters $\mathcal{E} = (x, y, a, b, \alpha)$ are the output of the process.

Importantly, the entire ellipse fitting is fully differentiable, as each step consists of differentiable operations. This enables gradient-based optimization, letting the parameters to be refined iteratively within an optimization framework.

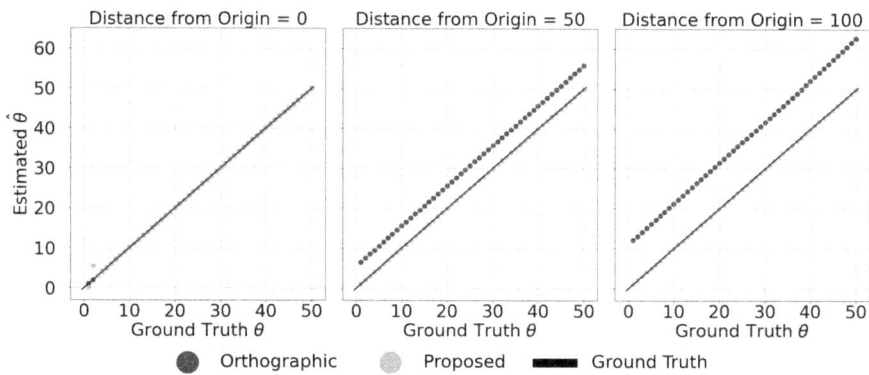

Fig. 2. Comparison of $\hat{\theta}$ estimation on proposed and orthographic projection [9] in the simulated environment. From left to right, experiments were conducted when distance from the origin to the object, (k, l), increases from 0, 50, and to 100 mm, while other parameters were fixed. The peak at distance = 0 occurs as $\theta \to 0$, which causes numerical instability as the ellipse collapses to a line.

2.2 Forward Model: 3D-Pose to the Image Plane

We start with landmark $\hat{S}$, which are points in a circle in a standard position, i.e., $(\theta, \varphi, k, l) = (0, 0, 0, 0)$, at a mean distance h along the extraplanar axis. The source is placed at height H above the detection plane. Importantly, by convention, $\theta = 0$ is orthogonal to the image plane ("zero ante/retroversion" in THA), so $x_i = r\cos\left(\frac{2\pi i}{n}\right), y_i = 0, z_i = h + r\sin\left(\frac{2\pi i}{n}\right)$ for n landmarks. The landmarks are rotated in 3D space by rotation matrix R:

$$R(\theta, \varphi) = \begin{pmatrix} \cos\varphi & -\sin\varphi & 0 \\ \sin\varphi & \cos\varphi & 0 \\ 0 & 0 & 1 \end{pmatrix} \begin{pmatrix} 1 & 0 & 0 \\ 0 & \cos\theta & -\sin\theta \\ 0 & \sin\theta & \cos\theta \end{pmatrix}. \quad (10)$$

By convention, we rotate by θ first, so as to avoid specifying the extra-planar rotation axis. The landmarks are then rotated and translated to $\hat{S}^T$, and projected to the 2D image plane:

$$\hat{S}^T = R(\hat{\theta}, \hat{\varphi})\hat{S} + \begin{pmatrix} \hat{k} \\ \hat{l} \\ 0 \end{pmatrix}, \qquad \hat{S}^P = P(\hat{S}^T) = \frac{H}{H - \hat{S}_z^T} \begin{pmatrix} \hat{S}_x^T \\ \hat{S}_y^T \\ 0 \end{pmatrix} \quad (11)$$

By using $\hat{S}^P$, we can fit another ellipse $\hat{E}$ with parameters $\hat{\mathcal{E}} = (\hat{x}, \hat{y}, \hat{a}, \hat{b}, \hat{\alpha})$.

Table 1. Quantitative results of pose estimation on the simulated environment (top) and Implant CT (bottom) were evaluated using Hausdorff Distance (HD), registration time, and mean absolute rotation and translation errors. Mem. indicates the maximum GPU memory required for 2D/3D registration on 640 × 640 images, with "–" for CPU-only cases. DIV marks where large in-plane spatial separations caused metric failure, as HD is highly sensitive to such differences.

Method	Mem.	HD	Time (s)	θ err (°)	φ err (°)	k, l err (mm)	h err (mm)
Orthographic [9]	–	–	–	5.46	2.18	–	–
Proposed	–	1.44	1.07	0.27	0.61	0.33	1.44
Intensity [6]	44 GB	71.02	139.78	19.79	78.38	21.20	208.12
Embedding [5]	42 GB	DIV	286.25	18.21	38.20	71.02	207.78
Proposed	–	6.91	1.21	1.53	2.16	1.12	9.76
Seg. + **Proposed**	0.4 GB	11.21	3.48	3.93	2.56	1.76	16.93

2.3 Loss Calculation and Optimization

We measure ellipse-to-ellipse distortion (Fig. 1(c)) using the mean squared error between parameter sets $\mathcal{E} = (x, y, a, b, \alpha)$ and the estimate $\hat{\mathcal{E}}$.

$$\mathcal{L} = \frac{1}{N} \sum_{i=1}^{N} (\mathcal{E}_i - \hat{\mathcal{E}}_i)^2, \qquad (12)$$

The error in the angular elements α are adjusted to account for angular periodicity, ensuring that discrepancies are correctly calculated by using $\min((\alpha - \hat{\alpha})^2, (180 - |\alpha - \hat{\alpha}|)^2)$. We then minimize the loss $\mathcal{L}$ by iteratively updating the parameters $(\hat{\theta}, \hat{\varphi}, \hat{k}, \hat{l}, \hat{h})$ using an analytical gradient computed through automatic differentiation (Fig. 1(d)), iterating until convergence.

3 Experiments

Experiments were conducted in three distinct environments: Numerical Simulation, Implant CT, and Cadaver CT. For each of these environments, the pose parameters were uniformly sampled for each trial point: Anteversion (θ) was sampled from the interval $(0, 50)$ degrees (based on the "safe zone" in implant anteversion [1,8]), inclination (φ) from $(-90, 90)$ degrees, in-plane translation (k, l) from $(-100, 100)$ mm, and extra-planar translation (h) from $(100, 520)$ mm. Projections onto the 2D plane were then simulated for the given pose parameters. All the experiments were conducted with source-to-detection distance (H) set to 1040 mm.

For optimization, post estimate variables were initialized to a "first best guess": $(\hat{\theta}_0, \hat{\varphi}_0, \hat{k}_0, \hat{l}_0, \hat{h}_0) = (25°, 40°, \beta E_x, \beta E_y, (1-\beta)H)$ where E_x and E_y is the center coordinates of the observed ellipse E, and β is the ratio between the radius of the implant r and major axis E_a of the observed ellipse.

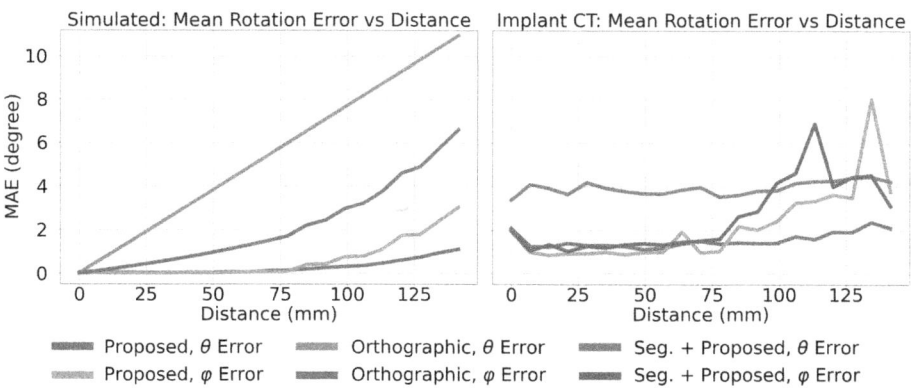

Fig. 3. Mean absolute error (MAE) of θ and φ as a function of the distance from the origin to the object, (k, l). Experiments were conducted in the synthetic environment (left) and the DRR environment (right). θ and φ indicate the experiments where each parameters were fixed. Results from intensity and embedding-based registration for implant CT was excluded due to high error.

For each of the environments, S^P construction differs: In the **Numerical Simulation** environment, landmarks S^P were directly projected to the image plane, i.e., no label or image noise was added for that experiment. For the **Implant CT** experiment environment, images were generated from CT of an implant outside of tissue using DRR methods. Two separate sub-experiments were conducted by extracting landmarks S^P either by manually annotating landmarks in the CT and projecting them to the image plane, or, reflecting a more realistic scenario, training a U-Net [11] to segment the ellipsoidal area in the 2D image, the boundary of which are used as landmarks S^P. In the **Cadaver CT** environment, 5 cadavers were given THA procedures and then imaged using CT. Simulated 2D imaging was then collected via DiffDRR, from which S^P were annotated. Each cadaver has both Left and Right joint replacements.

We use the Numerical Simulation environment to validate the perspective projection over the orthographic projection methods [9]. For the Implant CT environments we compare the performance of the proposed method to current 2D/3D image registration methods, either intensity-based [6] or embedding/feature based [5]. We use the cadaver CT to demonstrate the viability of the proposed method in imaging conditions with real tissue.

4 Results

Numerical Simulation: As shown in Table 1, our method achieved θ and φ MAE of 0.27 and 0.61 degrees, respectively. In contrast, orthographic projection had θ and φ errors of 5.46 and 2.18 degrees. Figure 2 further illustrates how orthographic projection caused increasing distortion, while the proposed method remained robust even as the implant moved away from the center.

Table 2. Quantative results on Cadaver CT using the proposed method showing the viability of the method in real surgical environments.

Cadaver	Implant	HD	Time (s)	θ err (°)	φ err (°)	k, l err (mm)	h err (mm)
C 1	L	2.18	1.08	0.92	0.22	0.48	3.33
	R	6.26	1.09	1.23	0.31	1.62	8.77
C 2	L	3.59	1.09	0.38	0.64	0.73	6.07
	R	0.24	1.12	0.52	1.37	0.65	1.42
C 3	L	2.58	1.08	0.25	0.97	0.58	4.63
	R	2.72	1.08	0.39	0.43	1.55	5.61
C 4	L	3.67	2.86	0.05	1.51	1.26	8.48
	R	2.22	1.08	0.28	0.84	1.56	5.86
C 5	L	6.94	1.07	0.46	0.31	0.56	6.28
	R	0.70	2.18	0.22	0.87	0.84	0.15
Mean		3.11	1.38	0.47	0.75	0.99	5.07

Implant CT: As shown in Table 1, the proposed method successfully registered the cup with an average HD of 6.91 pixels in just 1.21 s. Other errors remained within 1.53 and 2.16 degrees, 1.12 and 9.76 mm for $(\hat{\theta}, \hat{\varphi}, (\hat{k}, \hat{l}), \hat{h})$. Incorporating a segmentation model led to slightly higher errors due to minor mis-segmentations, but still outperformed other 2D/3D registration methods which failed to register correctly while consuming up to 287 s. The robustness of our method is also shown in Fig. 3, where estimation errors remain consistently low across varying distances. Moreover, this method operates entirely on the CPU, and even with segmentation, it only requires 0.4 GB of GPU memory, making it practical in the operating room with minimal hardware requirements.

Cadaver CT: As shown in Table 2, our method successfully registered to implants in cadaver data, achieving a mean HD of 3.17 pixels with an average registration time of 1.38 s. The improved performance on cadaver data compared to implant images is due to the clear visibility of ellipses in the DRRs of implants in the cadaver. In contrast, the implant CT environment include cases where ellipses are barely visible, leading to higher HD values and errors.

5 Discussion and Conclusion

This work presents an efficient and precise method for hip implant orientation estimation in THA, proving it as an advantageous and important special case of the broader inverse problem in projective geometry. While general solutions, such as those by Uneri et al. [15], require iterative forward estimations, our approach analytically solves the forward operation, significantly improving computational efficiency. Similarly, landmark registration methods [7] often rely on explicit correspondences between landmarks, which we solve by applying ellipse geometry instead of imposing artificial correspondences as in ICP methods [3].

By integrating perspective projection and ellipse fitting from landmarks, our method reduces computational inefficiency, achieving near real-time processing while mitigating projection distortions, particularly as the implant gets further from the image center. The high accuracy demonstrated in test cases on cadavers confirm the robustness of this approach in real surgical conditions. Furthermore, the differentiability of the ellipse fitting process allows for its integration into learning-based frameworks, such as deep neural networks. This facilitates the incorporation of data-driven refinement strategies, potentially enhancing pose estimation accuracy through end-to-end optimization.

Acknowledgments. This work was supported in part by NSF 2321684 and a VISE Seed Grant.

Disclosure of Interests. The authors have no competing interests to declare that are relevant to the content of this article.

References

1. Abdel, M.P., von Roth, P., Jennings, M.T., Hanssen, A.D., Pagnano, M.W.: What safe zone? The vast majority of dislocated THAs are within the Lewinnek safe zone for acetabular component position. Clin. Orthop. Related Res.® **474**(2), 386–391 (2015). https://doi.org/10.1007/s11999-015-4432-5
2. Anger, M., et al.: Prospect guideline for total hip arthroplasty: a systematic review and procedure-specific postoperative pain management recommendations. Anaesthesia **76**(8), 1082–1097 (2021)
3. Besl, P.J., McKay, N.D.: Method for registration of 3-d shapes. In: Sensor Fusion IV: Control Paradigms and Data Structures, vol. 1611, pp. 586–606. Spie (1992)
4. Fitzgibbon, A., Pilu, M., Fisher, R.: Direct least square fitting of ellipses. IEEE Trans. Pattern Anal. Mach. Intell. **21**(5), 476–480 (1999). https://doi.org/10.1109/34.765658
5. Gao, C., et al.: Generalizing spatial transformers to projective geometry with applications to 2D/3D registration. In: Martel, A.L., et al. (eds.) MICCAI 2020. LNCS, vol. 12263, pp. 329–339. Springer, Cham (2020). https://doi.org/10.1007/978-3-030-59716-0_32
6. Gopalakrishnan, V., Golland, P.: Fast auto-differentiable digitally reconstructed radiographs for solving inverse problems in intraoperative imaging. In: Chen, Y., et al. (eds.) Clinical Image-Based Procedures. CLIP 2022. LNCS, vol. 13746, pp. 1–11. Springer, Cham (2023). https://doi.org/10.1007/978-3-031-23179-7_1
7. Grupp, R.B., et al.: Automatic annotation of hip anatomy in fluoroscopy for robust and efficient 2D/3D registration. Int. J. Comput. Assist. Radiol. Surg. **15**(5), 759–769 (2020). https://doi.org/10.1007/s11548-020-02162-7
8. Lewinnek, G.E., Lewis, J.L., Tarr, R., Compere, C.L., Zimmerman, J.R.: Dislocations after total hip-replacement arthroplasties. J. Bone Joint Surg. **60**(2), 217–220 (1978). https://doi.org/10.2106/00004623-197860020-00014
9. Liaw, C.K., Hou, S.M., Yang, R.S., Wu, T.Y., Fuh, C.S.: A new tool for measuring cup orientation in total hip arthroplasties from plain radiographs. Clin. Orthop. Relat. Res. **451**, 134–139 (2006). https://doi.org/10.1097/01.blo.0000223988.41776.fa

10. O'Connor, P.B., et al.: The impact of functional combined anteversion on hip range of motion: a new optimal zone to reduce risk of impingement in total hip arthroplasty. Bone Joint Open **2**(10), 834–841 (2021)
11. Ronneberger, O., Fischer, P., Brox, T.: U-Net: convolutional networks for biomedical image segmentation. In: Navab, N., Hornegger, J., Wells, W.M., Frangi, A.F. (eds.) MICCAI 2015. LNCS, vol. 9351, pp. 234–241. Springer, Cham (2015). https://doi.org/10.1007/978-3-319-24574-4_28
12. Streck, L.E., Boettner, F.: Achieving precise cup positioning in direct anterior total hip arthroplasty: a narrative review. Medicina **59**(2), 271 (2023)
13. Suh, Y., Mika, A., Martin, J.R., Moyer, D.: Dilation-erosion methods for radiograph annotation in total knee replacement. In: Medical Imaging with Deep Learning, short paper track (2023)
14. Unberath, M., et al.: DeepDRR – a catalyst for machine learning in fluoroscopy-guided procedures. In: Frangi, A., Schnabel, J., Davatzikos, C., Alberola-López, C., Fichtinger, G. (eds.) Medical Image Computing and Computer Assisted Intervention – MICCAI 2018. MICCAI 2018. LNCS, vol. 11073, pp. 98–106. Springer, Cham (2018). https://doi.org/10.1007/978-3-030-00937-3_12
15. Uneri, A., et al.: Known-component 3D-2D registration for image guidance and quality assurance in spine surgery pedicle screw placement. In: III, R.J.W., Yaniv, Z.R. (eds.) Medical Imaging 2015: Image-Guided Procedures, Robotic Interventions, and Modeling, vol. 9415, p. 94151F. International Society for Optics and Photonics, SPIE (2015). https://doi.org/10.1117/12.2082210

ECG Report Generation with Diagnostic Knowledge Enhanced Prompt Learning

Panpan Fan[1], Xiaodong Yue[1(✉)], Yufei Chen[2], Zhipeng Wei[3], Jie Shi[3], and Zhikang Xu[4]

[1] Artificial Intelligence Institute, Shanghai University, Shanghai, China
yswantfly@shu.edu.cn
[2] School of Computer Science and Technology, Tongji University, Shanghai, China
[3] School of Computer Engineering and Science, Shanghai University, Shanghai, China
[4] Institute of Intelligent Information Processing, Shanxi University, Taiyuan, China

Abstract. Due to the complexity of 12-lead signals, the diversity of cardiac conditions, and the semantic gap between waveforms and clinical language, existing large language model (LLM)-based electrocardiogram (ECG) report generation remains challenging. In this paper, we propose a novel approach to ECG report generation through diagnostic knowledge-enhanced prompt learning. Specifically, a knowledge-aware module is constructed to extract waveform features and diagnostic cues via multilabel classification. These clinical semantic features are then fused with textual descriptions to form input prompts, thereby enhancing the semantic richness of the prompts. Moreover, by incorporating signal augmentation to capture fine-grained waveform semantics, we perform both intra-modal and cross-modal alignment between high-dimensional ECG signals and the generated text during model training, thereby improving the accuracy and relevance of report generation. A constraint-aware loss is further introduced to ensure the inclusion of essential diagnostic elements. Experiments on benchmark datasets demonstrate that our proposed method achieves superior performance on natural language generation metrics (e.g. BLEU, CIDEr-D), validating its effectiveness in both generation accuracy and clinical relevance. The code is available at: https://github.com/pangpanqiqi/ECG-Report-Prompt.

Keywords: ECG Report Generation · Clinical Knowledge Integration · Prompt Learning

1 Introduction

ECG is a widely used, non-invasive modality for assessing cardiac electrical activity and plays a central role in the early detection of cardiovascular diseases (CVDs), the leading cause of global mortality [5]. The growing volume of ECG exams, combined with a shortage of trained cardiologists, has strained manual interpretation workflows, leading to diagnostic delays, misinterpretations, and reporting backlogs, especially in resource-constrained settings [6,7]. To address these challenges, recent efforts have explored automated ECG report generation from raw signals, aiming to improve efficiency and consistency [17,22]. Robust, scalable ECG reporting systems are thus essential for advancing timely and equitable cardiovascular care [17].

ECG report generation aims to translate complex temporal waveforms into concise clinical descriptions, yet remains significantly underexplored compared to classification tasks such as arrhythmia detection or risk prediction [15,16]. Early approaches primarily relied on multi-task learning or template-based methods [20], often prioritizing linguistic fluency over diagnostic accuracy. More recent efforts have introduced large language models (LLMs) for end-to-end generation [12,22], leveraging domain-specific prompts or retrieval mechanisms. However, these methods struggle to capture subtle waveform abnormalities and often fail to generalize across diverse ECG patterns, leading to incomplete or clinically inaccurate reports. In parallel, vision-language large models (VLLMs) [8,9,13,24] and medical image report generators [2,3,10,21,23] have shown the potential of prompt tuning and cross-modal alignment, but are limited to static images. Existing ECG generation frameworks thus lack temporal sensitivity, reliability estimation, and interpretability. As illustrated in Fig. 1, generic language models often fail to capture rare or subtle ECG abnormalities, resulting in unreliable reports.

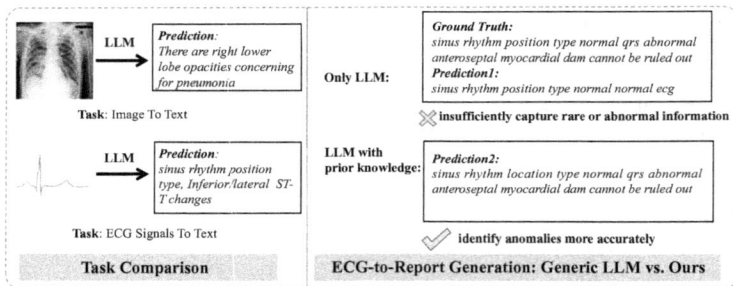

Fig. 1. Comparison between image-to-text and signal-to-text generation tasks and between a generic language model and our method, which leverages clinical knowledge to better capture rare ECG patterns and improve accuracy.

To address these challenges, we propose a novel generation framework that is clinically grounded and capable of generating coherent and diagnostically faithful ECG reports. Our framework addresses two key problems. First, we bridge the modality gap using cross-modal and intra-modal alignment strategies, with contrastive learning and signal augmentation to capture fine-grained waveform semantics. Second, we simulate the physician's diagnostic process through a dynamic prompting mechanism that integrates multi-label classification and signal-derived metrics (e.g. heart rate, PR interval, QRS duration), guided by clinical guidelines. Moreover, a constraint-aware loss penalizes the omission of

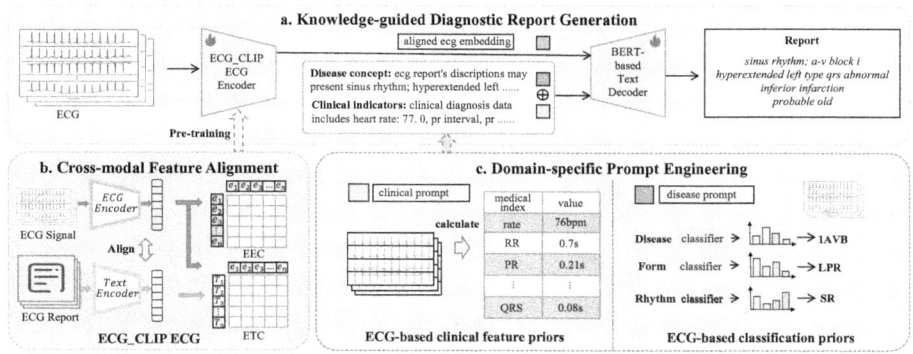

Fig. 2. Overview of the proposed framework. ECG signals are first encoded into aligned feature vectors via an ECG encoder. Clinical indicators and diagnostic labels are subsequently extracted and reformulated as textual prompts. A text decoder then generates diagnostic reports guided by both the ECG features and the clinical prompts.

essential diagnostic terms, enhancing report completeness and clinical reliability.
Our main contributions are:

1. **Physician-Workflow-Aligned Generation:** Propose a novel ECG report generation framework that mirrors cardiologists' diagnostic workflows by combining signal-based feature extraction with LLM-guided reasoning.
2. **Clinical-Knowledge-Fused Prompting:** Propose a prompting mechanism that integrates clinical priors and multi-label classification with constraint-aware loss to ensure inclusion of core diagnostic components.

2 Methods

In this section, we give a detailed introduction of the proposed framework. The model consists of three main components: ECG-to-report feature alignment for modality bridging, clinical knowledge-aware prompt construction, and cross-attention report generation, as illustrated in Fig. 2.

2.1 ECG-to-Report Feature Alignment for Modality Bridging

Cross-Modal Alignment. To bridge the modality gap between ECG signals and textual reports, we adopt a cross-modal alignment strategy. A ResNet1D encoder extracts temporal features from ECG signals $\mathbf{X}_e \in \mathbb{R}^{T \times D}$, while Med_CPT [4] encodes textual reports $\mathbf{X}_t \in \mathbb{R}^{L \times E}$. As shown in Fig. 2, both are projected into a shared embedding space via nonlinear mappings, yielding $\mathbf{z}_e, \mathbf{z}_t \in \mathbb{R}^d$. We employ a symmetric contrastive loss to align ECG-text pairs by maximizing similarity between matched representations and minimizing it for mismatched ones

$$\mathcal{L}_{\text{etc.}} = \frac{1}{2}(\mathcal{L}_{\text{e-t}} + \mathcal{L}_{\text{t-e}}), \quad (1)$$

where

$$\mathcal{L}_{\text{e-t}} = -\frac{1}{B}\sum_{i=1}^{B} \log \frac{\exp\left(\text{sim}(\mathbf{z}_{\text{e},i}, \mathbf{z}_{\text{t},i})/\tau\right)}{\sum_{j=1}^{B}\exp\left(\text{sim}(\mathbf{z}_{\text{e},i}, \mathbf{z}_{\text{t},j})/\tau\right)}, \quad (2)$$

$$\mathcal{L}_{\text{t-e}} = -\frac{1}{B}\sum_{i=1}^{B} \log \frac{\exp\left(\text{sim}(\mathbf{z}_{\text{t},i}, \mathbf{z}_{\text{e},i})/\tau\right)}{\sum_{j=1}^{B}\exp\left(\text{sim}(\mathbf{z}_{\text{t},i}, \mathbf{z}_{\text{e},j})/\tau\right)}. \quad (3)$$

where $\text{sim}(\cdot, \cdot)$ denotes cosine similarity, τ is a temperature parameter, and B is the batch size.

Intra-Modal Alignment. To enhance the robustness of ECG representations, we apply intra-modal contrastive learning, as shown in the EEC branch of Fig. 2. To enhance temporal diversity, we employ signal-level augmentations such as baseline drift simulation, random segment cutting, signal mixing, and temporal masking. Positive pairs are created by applying two independent dropout masks to the same signal embedding $\mathbf{z}_{\text{e},i}$

$$\mathbf{z}_{\text{e},i}^1 = \mathbf{z}_{\text{e},i} \odot \mathbf{M}^1, \quad \mathbf{z}_{\text{e},i}^2 = \mathbf{z}_{\text{e},i} \odot \mathbf{M}^2, \quad (4)$$

where $\mathbf{M}^1, \mathbf{M}^2$ are independent binary masks drawn from Bernoulli(p), with $p = 0.1$. The element-wise multiplication ($\odot$) generates two stochastic views of the same embedding. The contrastive loss encourages alignment between the two masked views while pushing apart different samples

$$\mathcal{L}_{\text{eec}} = -\frac{1}{B}\sum_{i=1}^{B} \log \frac{\exp\left(\text{sim}(\mathbf{z}_{\text{e},i}^1, \mathbf{z}_{\text{e},i}^2)/\tau\right)}{\sum_{j=1}^{B}\exp\left(\text{sim}(\mathbf{z}_{\text{e},i}^1, \mathbf{z}_{\text{e},j}^2)/\tau\right)}. \quad (5)$$

2.2 Clinical Knowledge-Aware Prompt Construction

To enhance interpretability, we construct structured prompts by coupling ECG biomarkers with clinical semantics (Fig. 2), including pathology classification and quantitative ECG analysis.

Pathology Classification Prompts. A ResNet1D-based multi-label classifier identifies diagnostic patterns from ECG signals $\mathbf{X}_e \in \mathbb{R}^{T \times D}$, producing label probabilities $\hat{y}_e \in \mathbb{R}^M$

$$\hat{y}_e^i = \sigma(\text{MLP}(\mathbf{E}_{\text{ResNet1D}}(\mathbf{X}_e))). \quad (6)$$

Each predicted label $\hat{y}_e^i$ is mapped to a standardized diagnostic term using a predefined mapping dictionary. For example, "AFIB" is converted to "Atrial Fibrillation". These mappings, derived from expert-defined templates, serve as semantic priors to guide the report generation process.

Quantitative Analysis Prompts. Numerical features, such as heart rate, RR interval, and PR interval, are extracted using NeuroKit through signal denoising and waveform segmentation. These metrics, combined with domain rules, are embedded as prompts to guide report generation. For instance, irregular RR intervals may suggest conduction disorders such as AV block.

2.3 Dynamically Constrained Report Generation

We generate diagnostic reports using a BERT decoder with causal self-attention and a cross-attention mechanism that incorporates ECG embeddings **X** and prompt features **D** (diseases) and **C** (clinical indicators).

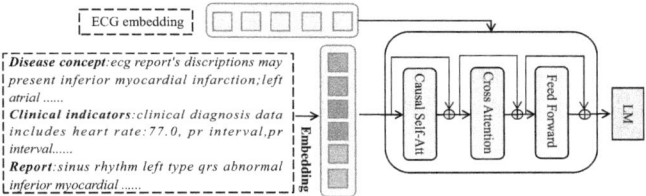

Fig. 3. Cross-attention module: Fuses ECG embeddings with prior prompts to enhance context-aware report generation.

At each decoding step, the model integrates both previously generated tokens and multimodal context to ensure clinical relevance. As illustrated in Fig. 3, the cross-attention module fuses ECG signals with structured prompts to enhance context-aware report generation. The training objective is defined as

$$\mathcal{L}_{\text{lm}} = -\sum_{t=1}^{T} \log p(y_t \mid y_{<t}, \mathbf{X}, \mathbf{D}, \mathbf{C}). \tag{7}$$

To promote inclusion of key diagnostic terms, we introduce a soft constraint loss. Given target terms $\mathcal{T} = \{t_1, \ldots, t_K\}$, we define

$$s_k = \max_{1 \leq j \leq L} p_j(t_k), \quad \mathcal{L}_{\text{constraint}} = -\frac{1}{K} \sum_{k=1}^{K} \log(s_k). \tag{8}$$

This differentiable loss encourages high token probabilities for essential clinical terms. The overall training objective combines alignment, generation, and constraint components

$$\mathcal{L}_{\text{total}} = \alpha \mathcal{L}_{\text{etc.}} + \beta \mathcal{L}_{\text{eec}} + \gamma (\mathcal{L}_{\text{lm}} + \frac{1}{2} \mathcal{L}_{\text{constraint}}). \tag{9}$$

The proposed method includes two training stages. In the alignment phase, the model is pre-trained on MIMIC-IV-ECG using a contrastive loss, where ECG signals are encoded by a ResNet1D network and textual reports by the Med-CPT model. In the generation phase, the model is fine-tuned on PTB-XL using a joint loss combining alignment and generation objectives. During inference, aligned ECG features are fused with prior prompts via a cross-attention mechanism, and the fused representation is decoded by a BERT-based decoder.

3 Experiments

We use MIMIC-IV-ECG dataset [1] and PTB-XL dataset [19]. After preprocessing, 771,693 ECGtext pairs were retained. PTB-XL dataset includes 21,837 10-second ECG recordings from 18,885 patients, each with a free-text report. It supports four multi-label tasks: Superclass, Subclass, Form, and Rhythm. We adopt the AdamW optimizer with an initial learning rate of 0.0001 and a batch size of 16. The hyperparameters α, β, and γ are set to 0.4, 0.1, and 0.5, respectively. To enhance generation quality and diversity, we set the temperature to 0.7, top-p to 0.8, and the repetition penalty to 1.0. All experiments are implemented in PyTorch and conducted on a single NVIDIA A100 GPU.

3.1 Comparison Study

Comparison with State-of-the-Art Methods. We compare our model with existing approaches using BLEU (1–4) [14], ROUGE-L [11], and CIDEr-D [18]. BLEU measures n-gram precision to assess fluency, ROUGE-L captures content overlap and sequence coherence, and CIDEr-D reflects clinical relevance based on TF-IDF weighted similarity. These metrics jointly evaluate the quality of generated ECG reports. As shown in Table 1, our method achieves superior performance across all metrics, notably in BLEU1 (0.432) and CIDEr-D (2.509), highlighting its ability to generate accurate and informative reports. The improved results can be attributed to our use of diagnostic labels to construct constraint-aware prompts. This guidance helps the model attend to clinically relevant information, enhancing both the fluency and diagnostic consistency of the generated reports.

Table 1. Performance comparison of different methods on six evaluation metrics

Method	BLEU1	BLEU2	BLEU3	BLEU4	ROUGE-L	CIDEr-D
Cross-modal Transfer (2023)	0.260	–	–	–	–	–
MEIT (GPT2-Medium) (2023)	0.329	0.278	0.254	0.232	0.391	2.120
ECG-Chat (2024)	0.323	–	–	0.110	0.300	–
Ours	**0.432**	**0.352**	**0.299**	**0.263**	**0.512**	**2.509**

Comparison of Language Model Backbones. We compare five backbones: BERT, GPT2-M, GPT2-L, LLaMA-2 (LoRA-tuned), and ours. ECG features are embedded into the first token, guided by a unified instruction prompt. ours incorporates structured clinical terms for fine-grained knowledge injection. Owing to its bidirectional nature, BERT better exploits diagnostic prompts, while autoregressive models (GPT, LLaMA) are less effective. Compared to autoregressive models, our work benefits from its bidirectional attention, which enables comprehensive understanding of the input prompt and downstream content. When guided by structured diagnostic terms, BERT can better align token-level representations with medical semantics, resulting in more precise and relevant outputs. As shown in Fig. 4, ours achieves the best overall performance. Qualitative results (Fig. 5 and Fig. 6) further show that it produces more detailed and clinically accurate reports compared to the vaguer outputs of other models.

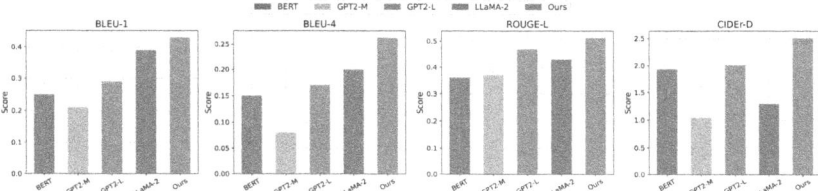

Fig. 4. Performance comparison of different language model backbones.

Generated Report:
GPT2-M: Sinus rhythm position type normal normal ecg ✗
GPT2-L: Sinus rhythm location type normal qrs abnormal anteroseptal infarction ✗
LLaMA-2: Sinus rhythm location type normal qrs abnormal anteroseptal infarction probable old inferior infarction probable old t abnormal in high lateral leads ✗
Ours: Sinus rhythm location type normal qrs abnormal anteroseptal myocardial dam cannot be ruled out ✓
Ground Truth: Sinus rhythm position type normal qrs abnormal anteroseptal myocardial dam cannot be ruled out

Fig. 5. Example outputs of ECG report generation from different models. Orange and cyan highlight descriptions are consistent with the ground truth. (Color figure online)

3.2 Ablation Study

Effect of Prompt and Constraint Loss on Report Quality. We evaluate three settings: no prompt, prompt only, and prompt with constraint loss. As shown in Fig. 7 (a), prompts significantly enhance performance by guiding generation, while the addition of constraint loss further improves factual accuracy and clinical relevance.

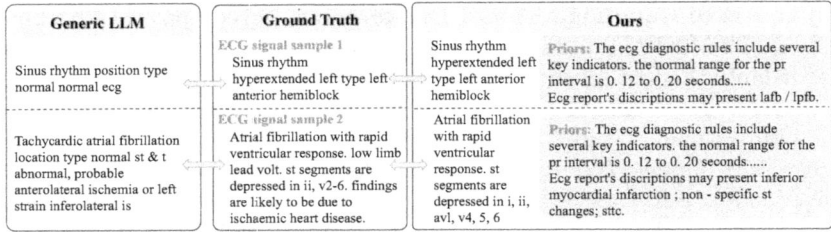

Fig. 6. Qualitative examples of the generic LLM and the proposed method. Green text indicates correct content generated by both models, while red text highlights correct details unique to ours. (Color figure online)

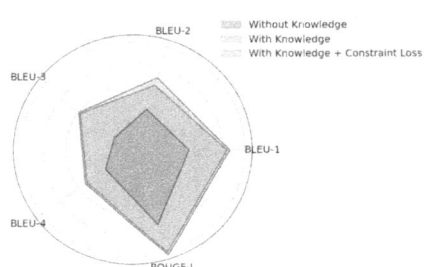
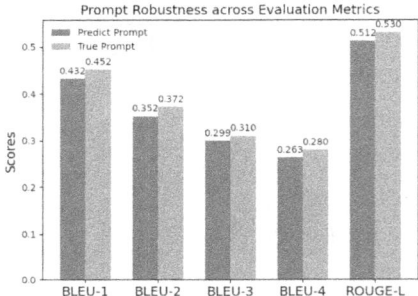

(a) Performance comparison under different configurations

(b) Performance comparison under various prompts

Fig. 7. Impact of model configuration and prompt design on diagnostic report generation.

Table 2. The classification performance on the PTB-XL dataset

Category	AUC (%)	Acc (%)	F1-Score (%)
all	**93.5**	**97.9**	72.1
sub_diag	91.6	96.5	67.4
super_diag	92.7	88.5	**75.8**
form	88.0	94.5	50.8
rhythm	90.4	89.9	49.8

Stability Analysis Under Label Perturbation. To incorporate structured prior knowledge, we perform multi-label classification on the PTB-XL dataset to extract 71 diagnostic, rhythm, and morphological labels using a ResNet1D model. Each classification task is evaluated independently using AUC, accuracy, and F1 score, as summarized in Table 2.

To further evaluate the stability of our framework under label perturbation, we compare report generation results using prompts constructed from either ground-truth or predicted diagnostic labels. As shown in Fig. 7 (b), the two settings yield highly similar outputs, demonstrating the robustness of our prompt construction strategy.

4 Conclusion

In this paper, we propose a prior-driven ECG report generation framework that integrates multi-modal alignment and clinical prompt engineering. By aligning temporal ECG features with textual semantics and introducing dynamic constraints, our method significantly improves the accuracy and clinical relevance of generated reports. Extensive experiments demonstrate consistent gains across BLEU and CIDEr-D metrics, validating the effectiveness and practical potential of our approach.

Despite promising results, this study remains at the algorithm validation stage. To assess generalization to out-of-distribution (OOD) data, we plan to conduct prospective clinical trials with partner hospitals. In addition, the current framework requires further optimization for real-time deployment. These findings highlight the potential of structured prior knowledge and prompt design to enhance the reliability and interpretability of automated report generation. In future work, we aim to further improve the computational efficiency of the model and expand its application scope to other medical domains, offering a more generalizable and scalable solution for multi-modal medical report generation.

Acknowledgments. This work was supported by the National Natural Science Foundation of China (Serial Nos. 62476165, 62406182) and Fundamental Research Program of Shanxi Province, China (Serial No. 202403021212176).

Disclosure of Interests. The authors have no competing interests to declare that are relevant to the content of this article.

References

1. Gow, B., et al.: Mimic-iv-ecg: diagnostic electrocardiogram matched subset. Type: dataset **6**, 13–14 (2023)
2. Huang, Z., Zhang, X., Zhang, S.: Kiut: knowledge-injected u-transformer for radiology report generation. In: Proceedings of the IEEE/CVF Conference on Computer Vision and Pattern Recognition, pp. 19809–19818 (2023)
3. Jin, H., Che, H., Lin, Y., Chen, H.: Promptmrg: diagnosis-driven prompts for medical report generation. In: Proceedings of the AAAI Conference on Artificial Intelligence, vol. 38, pp. 2607–2615 (2024)
4. Jin, Q., et al.: Medcpt: contrastive pre-trained transformers with large-scale pubmed search logs for zero-shot biomedical information retrieval. Bioinformatics **39**(11), btad651 (2023)
5. Johnson, L., et al.: Artificial intelligence for direct-to-physician reporting of ambulatory electrocardiography. Nat. Med. **31**(3), 925–931 (2025)
6. Kraik, K., et al.: The most common errors in automatic ecg interpretation. Front. Physiol. **16**, 1590170 (2025)

7. Lee, E., et al.: Artificial intelligence-enabled ecg for left ventricular diastolic function and filling pressure. NPJ Digit. Med. **7**(1), 4 (2024)
8. Li, J., Li, D., Savarese, S., Hoi, S.: Blip-2: bootstrapping language-image pre-training with frozen image encoders and large language models. In: International Conference on Machine Learning, pp. 19730–19742. PMLR (2023)
9. Li, J., Li, D., Xiong, C., Hoi, S.: Blip: bootstrapping language-image pre-training for unified vision-language understanding and generation. In: International Conference on Machine Learning, pp. 12888–12900. PMLR (2022)
10. Li, M., Lin, B., Chen, Z., Lin, H., Liang, X., Chang, X.: Dynamic graph enhanced contrastive learning for chest x-ray report generation. In: Proceedings of the IEEE/CVF Conference on Computer Vision and Pattern Recognition, pp. 3334–3343 (2023)
11. Lin, C.Y.: Rouge: a package for automatic evaluation of summaries. In: Text Summarization Branches Out, pp. 74–81 (2004)
12. Liu, C., Ma, Y., Kothur, K., Nikpour, A., Kavehei, O.: Biosignal copilot: leveraging the power of llms in drafting reports for biomedical signals. medrxiv (2023)
13. Liu, H., Li, C., Wu, Q., Lee, Y.J.: Visual instruction tuning. Adv. Neural Inf. Process. Syst. **36** (2024)
14. Papineni, K., Roukos, S., Ward, T., Zhu, W.J.: Bleu: a method for automatic evaluation of machine translation. In: Proceedings of the 40th Annual Meeting of the Association for Computational Linguistics, pp. 311–318 (2002)
15. Qiu, J., et al.: Transfer knowledge from natural language to electrocardiography: can we detect cardiovascular disease through language models? arXiv preprint arXiv:2301.09017 (2023)
16. Qiu, J., et al.: Automated cardiovascular record retrieval by multimodal learning between electrocardiogram and clinical report. In: Machine Learning for Health (ML4H), pp. 480–497. PMLR (2023)
17. Tang, J., Xia, T., Lu, Y., Mascolo, C., Saeed, A.: Electrocardiogram report generation and question answering via retrieval-augmented self-supervised modeling. arXiv preprint arXiv:2409.08788 (2024)
18. Vedantam, R., Lawrence Zitnick, C., Parikh, D.: Cider: consensus-based image description evaluation. In: Proceedings of the IEEE Conference on Computer Vision and Pattern Recognition, pp. 4566–4575 (2015)
19. Wagner, P., et al.: Ptb-xl, a large publicly available electrocardiography dataset. Scientific Data **7**(1), 1–15 (2020)
20. Yan, B., Pei, M.: Clinical-bert: vision-language pre-training for radiograph diagnosis and reports generation. In: Proceedings of the AAAI Conference on Artificial Intelligence, vol. 36, pp. 2982–2990 (2022)
21. Yang, S., Wu, X., Ge, S., Zheng, Z., Zhou, S.K., Xiao, L.: Radiology report generation with a learned knowledge base and multi-modal alignment. Med. Image Anal. **86**, 102798 (2023)
22. Yu, H., Guo, P., Sano, A.: Zero-shot ecg diagnosis with large language models and retrieval-augmented generation. In: Machine Learning for Health (ML4H), pp. 650–663. PMLR (2023)
23. Zheng, L., et al.: Judging llm-as-a-judge with mt-bench and chatbot arena. Adv. Neural. Inf. Process. Syst. **36**, 46595–46623 (2023)
24. Zhu, D., Chen, J., Shen, X., Li, X., Elhoseiny, M.: Minigpt-4: enhancing vision-language understanding with advanced large language models. arXiv preprint arXiv:2304.10592 (2023)

TileDVP: Decoding the Tissue Proteome from H&E Images

Émilie Mathian[1], Lukas Oldenburg[1], Eduard Chelebian[1], Lisa Schweizer[1], Gijs Zonderland[1], Kristian Egebjerg[2], Ishani Ummat[1], Andreas Mund[1], and Maximillian T. Strauss[1] (✉)

[1] OmicVision Biosciences, Copenhagen, Denmark
mstrauss@omicvision.com
[2] Department of Oncology, Copenhagen University Hospital, Rigshospitalet, Copenhagen, Denmark
https://www.omicvision.com/

Abstract. Recent advances in multimodal deep learning have successfully begun to link molecular data with tissue morphology. To date, this work has largely focused on transcriptomics, a limited surrogate for protein expression. The direct prediction of proteins, the functional endpoints of gene expression, remains underexplored, largely due to the difficulty in generating spatially resolved proteomic data. Early efforts to predict proteomic data from tissue morphology were hindered by low specificity and limited multiplexing capacity, constraints inherent to immunohistochemistry and multiplexed immunofluorescence techniques. This proof-of-concept study presents a novel data generation pipeline to address these throughput and specificity challenges and decode the proteome from H&E slides via high-throughput tile-level mass spectrometry profiling. This approach enables a direct one-to-one correspondence between histological features and proteomic measurements, which is an essential prerequisite for training robust foundation models.

By applying this pipeline to gastric cancer biopsies, the study demonstrated that morphologically distinct clusters corresponded to distinct proteomic profiles. Notably, tumor regions were enriched for clinically relevant markers such as HMGB1, LGALS3, and ERBB2, all of which are associated with poor prognosis.

This work establishes the foundation for a new generation of AI-driven proteomics, demonstrating that routine histological images contain sufficient information to predict thousands of proteins across diverse biological conditions. TileDVP represents a paradigm shift toward accessible, high-throughput spatial proteomics that could transform biomarker discovery and precision medicine applications.

Keywords: Deep Visual Proteomics · Deep Learning · Histopathology

E. Mathian and L. Oldenburg—These authors contributed equally.

Supplementary Information The online version contains supplementary material available at https://doi.org/10.1007/978-3-032-05479-1_3.

1 Introduction

Recent advances in multimodal deep learning have demonstrated the feasibility of integrating transcriptomic and spatial transcriptomics data with morphological features extracted from hematoxylin and eosin (H&E)-stained whole slide images (WSIs). These methods demonstrate that morphological features can partially predict RNA-sequencing signatures. While advances in genomics and transcriptomics have significantly expanded our molecular understanding of disease, proteins–which represent the functional endpoint of gene expression and are the most clinically actionable molecular markers-remain underexplored as targets for direct prediction from tissue morphology in the deep learning community.

Immunohistochemistry (IHC), the standard method for assessing protein expression, is limited by interpretative variability and low multiplexing capacity. In gastric cancer, for example, HER2 protein, encoded by the ERBB2 gene, is a key prognostic and predictive marker; however, IHC shows poor agreement in equivocal (2+) cases due to inter-observer variability and staining heterogeneity [18]. To overcome these limitations, mass spectrometry (MS)-based spatial proteomics offers a multiplexed and quantitatively robust alternative. In particular, Deep Visual Proteomics (DVP) allows for high-depth, spatially resolved protein profiling and robust biomarker discovery across thousands of targets, offering a powerful new avenue for precision medicine applications [8,9,12,15].

We present a scalable proof-of-concept pipeline that directly links H&E tissue morphology to spatially resolved proteomic data through DVP. Applied to gastric cancer biopsies, this approach yielded over 600 tile-level measurements with one-to-one correspondence to histological features, offering a high-depth alternative to IHC, multiplexed immunofluorescence (MIF), and RNA-based inference by targeting proteins–the direct and druggable effectors of cellular function (Fig. 1). The main contributions of this work are fourfold: (i) we present a generalizable pipeline for MS-based spatial proteomics, termed TileDVP, a derivative of the DVP approach; (ii) we introduce an AI-based tile selection strategy that maximizes morphological diversity; (iii) we provide a foundation for cross-modal analysis between the proteome and H&E tile morphology, including predictive modeling and intra-tumor heterogeneity assessment; and (iv) we investigate which proteins are predictable from histology, highlighting the challenges of mapping protein abundance onto tissue architecture.

2 Related Work

2.1 Predicting Gene Expression from H&E

Pioneering models, such as ST-Net [5] and Path2Space [17], demonstrated that simple CNN or MLP architectures can predict RNA-seq profiles from H&E morphology, showing that transcriptomic signals are partially encoded in tissue architecture. However, most models were trained on small cohorts, often fewer than 25 patients, which limits generalizability [2]. Importantly, the field is now transitioning from shallow encoders to spatially aware architectures, such as

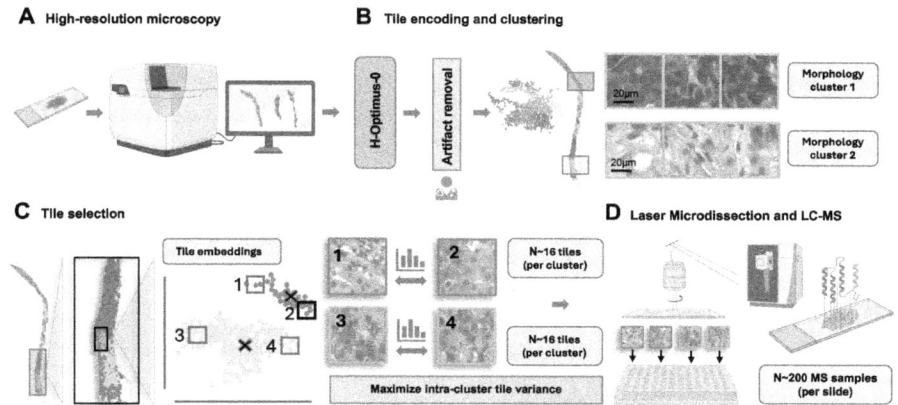

Fig. 1. Tile-level Deep Visual Proteomics workflow. A) H&E-stained tissue sections are imaged. **B)** Digitized images are tiled and embedded in the H-Optimus-0 latent space, where unsupervised clustering finds morphological groups per patient. **C)** From each slide, approximately 200 representative tiles are selected to cover a variety of morphologies, while maximizing intra-cluster variance. **D)** The corresponding tile-shaped tissue regions are isolated by laser-capture microdissection, and proteomes from 600 tiles were acquired with an Orbitrap Astral mass spectrometer.

Vision Transformers [3] and Graph Neural Networks [4,14]. These more complex models could benefit from larger and more diverse datasets, such as HEST-1K [6]. Extending predictions to proteins, given their link to tissue function and clinical outcomes, and developing scalable pipelines for spatial proteomics histology data is the logical next step.

2.2 Predicting Proteins from H&E Stained Tissue

Among the few studies specifically targeting spatial proteomics, DeepS4P [11] proposed a deep learning-guided sparse sampling strategy to reconstruct proteomic maps from limited MS measurements. DeepS4P [11] employs an untargeted grid-based sampling method. By sampling parallel tissue strips and applying a multilayer perceptron to infer unsampled regions, DeepS4P demonstrates how deep learning can expand sparse proteomic coverage. However, this untargeted approach does not incorporate morphological features of tissue to guide sampling.

2.3 Evaluation Metrics

Performance reporting in spatial omics also faces methodological challenges. Most models report the Pearson correlation coefficient (PCC) on highly expressed genes, which inflates performance by favoring easier-to-predict targets and often overlooks biologically significant extreme values–the outliers that fuel biological discovery–which tend to be under-predicted and have little impact on global PCC values. Instead of focusing on highly expressed genes, MISO [14] proposed using spatial autocorrelation as a more biologically meaningful criterion. Continuing in this direction, we investigated which proteins are more tractable to predict from H&E-stained tissue, acknowledging that not all proteins are independently correlated with distinct morphological features.

These methodological limitations highlight the need for a new pipeline capable of generating spatially resolved, high-depth proteomic data, as enabled by TileDVP.

3 Materials and Methods

3.1 Overview of Deep-Visual-Proteomics

DVP, a spatial proteomics method, integrates high-resolution histology, laser microdissection, and ultra-sensitive MS to generate spatially resolved proteomic maps. In this study, we introduce TileDVP, a variant of DVP that profiles proteins on individual excised tissue squares. Instead of operating at the cell phenotype level, this adjustment enables a direct, localized comparison between image-level histological features and protein composition and thus directly supports the use of foundation model embeddings. In our biopsy-section experiments, TileDVP preserved the proteomic depth of traditional DVP, quantifying an average of 3,989 proteins per tile (Table S1) [8], across approximately 200 tiles measured per patient, and maintains a broad spatial coverage (Fig. 1).

3.2 Patient Cohort and Imaging

Three gastric adenocarcinoma core needle biopsies (S001, S005, S008) were collected at Copenhagen University Hospital, Rigshospitalet. For all samples, ERBB2 status was confirmed by IHC, given its clinical relevance for prognosis and targeted treatment with trastuzumab [18] (see Supplementary Methods).

3.3 Morphology-Guided Tile Selection

For computational analysis and after background removal, tissue sections were processed into non-overlapping tiles (55×55 μm at 0.19 mpp) to retain essential cellular-scale morphological information and context without compromising the integrity of the proteomic signal. Morphological embeddings were computed using H-Optimus-0, a ViT-B/16 foundation model trained on over 10 million histology tiles [13]. The last layer of this model produces a 1,536-dimensional normalized embedding vector for each tile, which served as the input for clustering, selection, and predictive modeling. We optimized MS sample selection to cover broad tissue diversity. To this end, embedding vectors from all three patients were independently clustered using K-means, initially generating 30 distinct morphological groups. After manual removal of artifactual clusters such as background and blurry regions, 12 biologically relevant clusters remained.

For each cluster, we applied outlier filtering based on tile-to-cluster-center distance. To maximize morphological diversity, we partitioned each cluster into a fixed number of quantiles based on each tile's distance from the cluster centroid, then sampled tiles from each quantile to maximize within-cluster variance. To minimize spatial redundancy, we enforced a minimum distance of 150 μm between any two selected tiles. Approximately 200 MS samples per patient (~16 tiles per cluster) were selected for downstream MS analysis. Overall, the tile selection strategy ensures a diverse representation of tissue morphology while maintaining spatial coverage across the tissue (Fig. 1).

3.4 Modeling Strategy and Evaluation Metrics

To quantify congruence between proteomic and morphological modalities, we projected KNN-imputed proteomic measurements onto a two-dimensional space computed with UMAP and overlaid morphological clusters obtained with k-means (k = 6). These clusters were computed from H-Optimus-0 embedding vectors of all acquired H&E tiles, across all patients, in contrast to previous clustering. The number of clusters (k) was set as small as possible to retain simple morphological groups, based on feedback from a pathologist. We then measured concordance as the ratio of inter- to intra-cluster distances (see Supplementary Methods).

To predict proteomic expression from histology, we used univariate XGBoost regression, leveraging its performance with scarce data and its ability to handle non-imputed MS measurements directly. Owing to pronounced inter-patient proteomic variability, we trained an independent regression model for each protein within each patient. Cross-patient modeling was not feasible unless the cohort size was substantially increased. XGBoost received H-Optimus-0 tile embeddings as input, without further external normalization. The output corresponds to the log2-normalized, experimentally measured protein abundance for each tile. Model performance was evaluated using the Pearson correlation coefficient (PCC), mean absolute error (MAE), and the coefficient of determination (R^2) between predicted and observed protein levels. We investigated whether a protein's spatial organization affects how well it can be predicted from morphology. We measured this organization using Moran's Index, where positive values indicate spatial organization and near-zero or negative values suggest random patterns. We hypothesized that proteins exhibiting positive spatial autocorrelation would be more amenable to morphology-based prediction.

4 Results

4.1 Quantitative Results

To assess the feasibility of predicting MS protein intensities from H&E-stained tiles, we examined the alignment between morphological and molecular identities. Our analysis revealed that tiles closely positioned in the molecular embedding space generally shared similar morphological characteristics. This concordance is visually represented by the shared color of neighboring points (Fig. 2A).

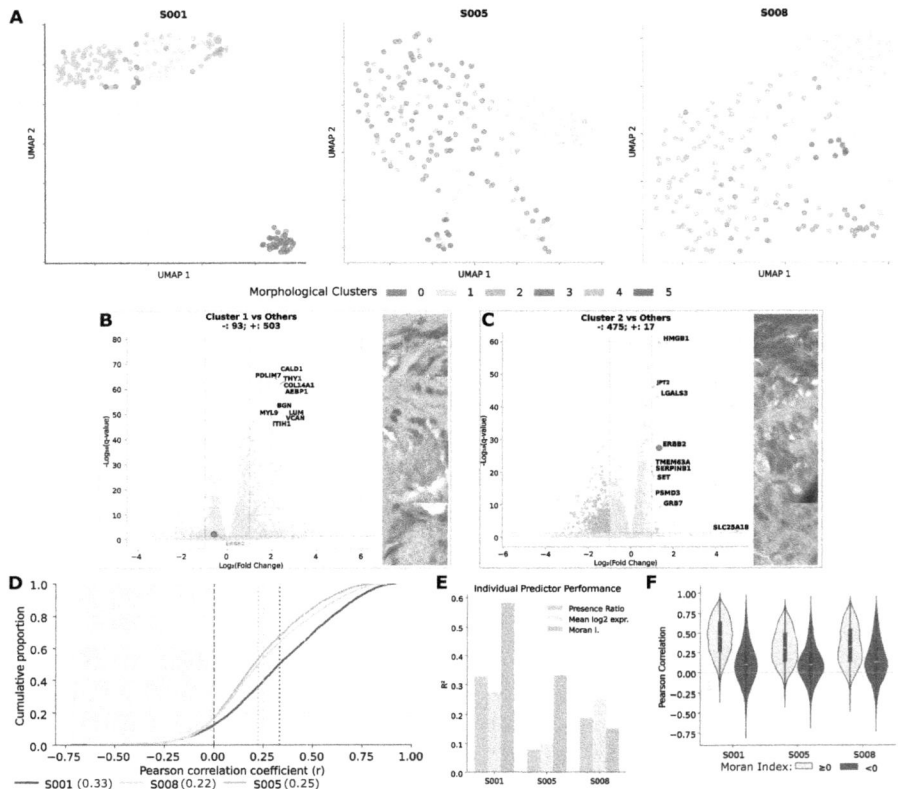

Fig. 2. Prediction of MS protein measurements from morphology. A) UMAP projections of imputed MS protein intensities for three gastric tumor samples. Points represent tile-level MS measurements colored by morphological clusters (k = 6, via K-means clustering). **B)** Volcano plot with $\log_2$ fold changes in protein abundance comparing cluster 1 (non-tumor) versus all other clusters. Highlighted points indicate proteins significantly differentially expressed (FDR < 0.05, —FC— > 2); counts of down- and up-regulated proteins are provided in the title. The ten most up-regulated proteins are labeled. A random selection of four tiles from cluster 1 are shown. **C)** Similar to B, differential expression analysis of tumoral Cluster 2 highlights ERBB2 overexpression ($FC = 2.5$) for reference. **D)** Cumulative distributions of PCC from univariate XGBoost classifiers trained individually per sample. Vertical dashed lines indicate mean PCC values. **E)** Coefficients of determination from univariate linear models predicting PCC based on: (i) protein presence ratio, (ii) mean $\log_2$ protein expression, and (iii) Moran's I spatial autocorrelation index. **F)** Violin plots comparing PCC distributions by protein spatial organization (positive vs. negative Moran's I).

Consequently, tiles within the same morphological cluster tend to group closely within the molecular embedding space, with an average inter-/intra-cluster distance ratio of 3.1 (range: 1.1–7.6), significantly higher than expected under random cluster assignments (permutation test, p < 0.005 for all samples).

The defined morphological clusters correspond to visually distinguishable morphologies (Figure S1). Clusters 3 and 4 demonstrated limited interpretability: cluster 3 primarily included out-of-focus tiles, and cluster 4 was exclusive to sample S001. Clusters 1 and 5 predominantly represented non-tumor areas, as evidenced by their lower expression of EPCAM [19]; in contrast, clusters 0 and 2 were identified as tumoral (Figure S2). These morphological clusters each present distinct protein expression profiles (Fig. 2B-C, S3, S4). Non-tumoral Cluster 1 exhibits overexpression of extracellular matrix and fiber-related proteins (Figure S4), typical of such regions. Conversely, cluster 2 comprises tumoral tiles (Figure S1) and is defined by a proteomic profile enriched in markers associated with poor prognosis in gastric cancer–such as HMGB1 [20], LGALS3 [10], SERPINB1 [7], and notably ERBB2 [1] (Fig. 2C). Figure 2D demonstrates that morphology alone provides a robust signal for proteomic prediction. Univariate models achieved a mean PCC of 0.27, a mean MAE of 0.11, and a mean R^2 of –0.09 across all proteins, with 40% of proteins exhibiting a positive R^2. For comparison, models trained on permuted data yielded a mean PCC of –0.02, a mean MAE of 0.13, and a mean R^2 of –0.39, with only 2.5% of proteins exhibiting a positive R^2. The obtained values are comparable to transcriptomic benchmarks–e.g., TRIPLEX reports median PCCs of 0.29 (all genes) and 0.45 (top 250 genes)–with the important caveat that our models are trained within-sample and not across patients [3]. As with transcriptomic data, we hypothesize that not all proteins are reflected in morphological features. Therefore, we investigated variability in protein predictability and confirmed that both the protein-presence ratio (i.e., the frequency of detection across tiles) and the expression level significantly influenced predictive performance. Consistent with MISO [14], we also found that proteins exhibiting stronger spatial organization, and thus likely expressed differentially across morphological entities, were associated with enhanced predictability (Fig. 2E–F). Notably, we do not expect mass spectrometry data to follow the same predictability profile as transcriptomics, given the differences in measurement dynamic range between the two technologies. Given ERBB2's therapeutic importance [1] and its role in morphological cluster 2 (Fig. 2C), we focused on predicting its expression.

4.2 Focus Around ERBB2

Our model performed well for samples S001 (PCC = 0.41) and S008 (PCC = 0.76), but performance was limited for sample S005 (PCC = 0.01), which exhibited low and sparse ERBB2 detection by MS (Fig. 3A, B). Intriguingly, the low MS signal and corresponding poor model performance for S005 mirrors its challenging clinical classification as an IHC 2+ (equivocal) case, which required subsequent fluorescence in situ hybridization (FISH) analysis to be confirmed as positive. This stands in sharp contrast to the definitive statuses of S001 (IHC 1+, negative) and S008 (IHC 3+, positive) (Figure S8) [18]. Qualitatively,

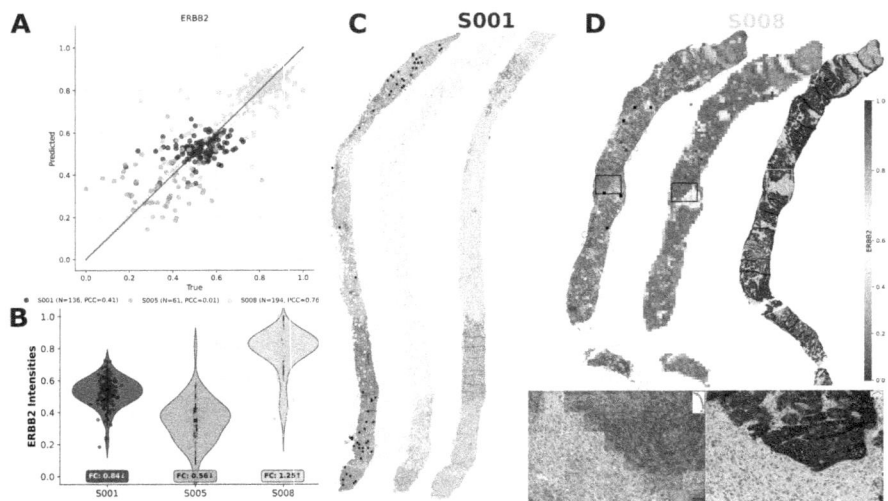

Fig. 3. Prediction of ERBB2 expression across samples. A) Scatterplot comparing predicted and measured ERBB2 expression values ($\log_2$-transformed and min-max scaled) for three different samples. Each sample was modeled independently, and PCC are reported in the legend. **B)** Violin plots of normalized ERBB2 intensity values as measured by MS for each sample. The FC values shown below each plot represent the ratio of the sample mean to the global mean ERBB2 intensity. **C)** Qualitative spatial analysis for sample S001. Left: H&E-stained WSI overlaid with tile-level ERBB2 measurements obtained by MS (blue = low, red = high, black = not detected). Center: prediction heatmap of normalized ERBB2 intensities using the same color scale. Right: corresponding ERBB2 IHC. **D)** Equivalent visual comparison for sample S008. Below, a magnified region highlights local agreement between ERBB2 expression, as predicted by tile color on the H&E overlay, and the corresponding IHC signal. (Color figure online)

extending our ERBB2 predictions to every tile in the WSIs provides a more comprehensive characterization of intra-tumor heterogeneity than the inherently sparse MS measurements. For sample S001 (IHC 1+, ERBB2 negative), tiles from morphological clusters 3 and 5 exhibit very low predicted ERBB2 intensity, in agreement with MS data (Fig. 3C, Figure S6). Conversely, for sample S008, the best-predicted case, regions predicted to have high or low ERBB2 expression demonstrate near-perfect alignment with the corresponding IHC stain (Fig. 3D). In contrast, the very low prediction performance for sample S005 prohibits meaningful interpretation of its WSI-level ERBB2 predictions (Figure S7). These qualitative observations regarding ERBB2 suggest that morphology-based predictions can significantly enhance our understanding of intra-tumor heterogeneity. This approach offers a promising avenue for other key proteins and also highlights the critical role of MS measurements in resolving clinically ambiguous cases.

5 Discussion and Perspectives

The foregoing results demonstrate that routine H&E morphology harbors sufficient information to predict thousands of spatially resolved protein abundances, establishing TileDVP as a practical bridge between histology and proteomics [8]. This proof-of-concept study highlights how morphologically distinct clusters correspond to unique proteomic profiles, for example, we identified a morphological cluster enriched for clinically relevant markers with prognostic value, including HMGB1, LGALS3, and ERBB2 [1,10,20].

Our analysis confirmed that proteins with low detection frequency, expression, or spatial organization were less predictable, an expected outcome, as not all proteins have strong morphological correlates. Nonetheless, future AI models leveraging MS-based proteomics could fully exploit the high sensitivity and depth of this technique, enabling access to thousands of proteins. This provides a powerful complement to approaches like the recent KRONOS experiment [16], which demonstrated advanced spatial tissue profiling capabilities with impressive multiplexing, yet remained limited in proteome coverage. Furthermore, because MS-based proteomics fundamentally identifies proteins based on individual peptides, it enables the direct detection of protein isoforms and post-translational modifications, largely inaccessible to antibody-based methods.

The data generation pipeline is highly extensible and paves the way for large-cohort, disease-agnostic studies. While the current cohort size precluded the use of spatially aware models such as Vision Transformers [3] or Graph Neural Networks [4,14], future large-scale datasets will enable their integration. These architectures, proven in spatial transcriptomics, will help capture complex tissue context. Further model refinement may incorporate contrastive learning to improve robustness and focus on differential expression that drives biological information. The generation of new data will help improve the generalizability of models across patients, particularly given the relative rather than absolute nature of protein abundance measurements in MS-based proteomics.

By scaling this data generation engine, we can build models that enhance the interpretability of even sparse MS data and support AI-guided pre-selection of molecularly relevant regions for targeted analysis. This work establishes the foundation for a transformative shift in how we approach spatial proteomics, moving from resource-intensive, specialized techniques to accessible, routine applications that can be integrated into standard clinical workflows. Future developments will focus on scaling to larger, more diverse cohorts and integrating advanced spatial modeling approaches to further enhance predictive performance and clinical utility.

Disclosure of Interests. All authors, except KE, are employees of OmicVision Biosciences. KE, MD, has received funding and support from pharmaceutical and diagnostic companies and serves as Chairman of the Danish Society for Cyto- and Histochemistry.

Ethical Approval. The study was conducted in accordance with the Declaration of Helsinki. Regulatory approvals were obtained from the Regional Ethics Committee and the Danish Data Protection Agency (Danish Ethical Committee, file number: 1300530). All patients signed informed written consent.

Data and Code Availability. The datasets and code used to generate the results presented in this study are available from the corresponding author upon request. We are pleased to share them for the purpose of reproducing our analyses or conducting further academic research.

References

1. Barros-Silva, J., et al.: Association of erbb2 gene status with histopathological parameters and disease-specific survival in gastric carcinoma patients. Br. J. Cancer **100**(3), 487–493 (2009)
2. Chelebian, E., Avenel, C., Wählby, C.: Combining spatial transcriptomics with tissue morphology. Nat. Commun. **16**(1), 1–13 (2025)
3. Chung, Y., Ha, J.H., Im, K.C., Lee, J.S.: Accurate spatial gene expression prediction by integrating multi-resolution features. In: Proceedings of the IEEE/CVF Conference on Computer Vision and Pattern Recognition, pp. 11591–11600 (2024)
4. Ganguly, A., et al.: Merge: multi-faceted hierarchical graph-based gnn for gene expression prediction from whole slide histopathology images. arXiv preprint arXiv:2412.02601 (2024)
5. He, B., et al.: Integrating spatial gene expression and breast tumour morphology via deep learning. Nat. Biomed. Eng. **4**(8), 827–834 (2020)
6. Jaume, G., et al.: Hest-1k: a dataset for spatial transcriptomics and histology image analysis. Adv. Neural Inf. Process. Syst. **37**, 53798–53833 (2024)
7. Kwon, C., et al.: Serpin peptidase inhibitor clade a member 1 is a biomarker of poor prognosis in gastric cancer. Br. J. Cancer **111**(10), 1993–2002 (2014)
8. Mund, A., et al.: Deep visual proteomics defines single-cell identity and heterogeneity. Nat. Biotechnol. **40**(8), 1231–1240 (2022)
9. Nordmann, T.M., et al.: Spatial proteomics identifies jaki as treatment for a lethal skin disease. Nature 1–9 (2024)
10. Okada, K., Shimura, T., Suehiro, T., Mochiki, E., Kuwano, H.: Reduced galectin-3 expression is an indicator of unfavorable prognosis in gastric cancer. Anticancer Res. **26**(2B), 1369–1376 (2006)
11. Qin, R., Ma, J., He, F., Qin, W.: In-depth and high-throughput spatial proteomics for whole-tissue slice profiling by deep learning-facilitated sparse sampling strategy. Cell Discovery **11**(1), 21 (2025)
12. Rosenberger, F.A., et al.: Spatial single-cell mass spectrometry defines zonation of the hepatocyte proteome. Nat. Methods **20**(10), 1530–1536 (2023)
13. Saillard, C., et al.: H-optimus-0. GitHub (2024). https://github.com/bioptimus/releases/tree/main/models/h-optimus/v0, gitHub repository
14. Schmauch, B., et al.: A deep learning-based multiscale integration of spatial omics with tumor morphology. bioRxiv, pp. 2024–07 (2024)
15. Schweizer, L., et al.: Spatial proteo-transcriptomic profiling reveals the molecular landscape of borderline ovarian tumors and their invasive progression. Cancer Cell (2023)

16. Shaban, M., et al.: A foundation model for spatial proteomics. arXiv preprint arXiv:2506.03373 (2025)
17. Shulman, E.D., et al.: Path2space: an ai approach for cancer biomarker discovery via histopathology inferred spatial transcriptomics. Cancer Res. **85**(8_Supplement_1), 6353 (2025)
18. WCoTE, B.: Digestive system tumours. Lyon (France): international agency for research on cancer (2019)
19. Wenqi, D., et al.: Epcam is overexpressed in gastric cancer and its downregulation suppresses proliferation of gastric cancer. J. Cancer Res. Clin. Oncol. **135**, 1277–1285 (2009)
20. Zhang, J., Kou, Y.B., Zhu, J.S., Chen, W.X., Li, S.: Knockdown of hmgb1 inhibits growth and invasion of gastric cancer cells through the nf-κb pathway in vitro and in vivo. Int. J. Oncol. **44**(4), 1268–1276 (2014)

Domain-Specialized Interactive Segmentation Framework for Meningioma Radiotherapy Planning

Junhyeok Lee[1], Han Jang[2], and Kyu Sung Choi[2](✉)

[1] Seoul National University of Medicine, Seoul, Republic of Korea
[2] Seoul National University Hospital, Seoul, Republic of Korea
ent1127@snu.ac.kr

Abstract. Precise delineation of meningiomas is crucial for effective radiotherapy (RT) planning, directly influencing treatment efficacy and preservation of adjacent healthy tissues. While automated deep learning approaches have demonstrated considerable potential, achieving consistently accurate clinical segmentation remains challenging due to tumor heterogeneity. Interactive Medical Image Segmentation (IMIS) addresses this challenge by integrating advanced AI techniques with clinical input. However, generic segmentation tools, despite widespread applicability, often lack the specificity required for clinically critical and disease-specific tasks like meningioma RT planning. To overcome these limitations, we introduce Interactive-MEN-RT, a dedicated IMIS tool specifically developed for clinician-assisted 3D meningioma segmentation in RT workflows. The system incorporates multiple clinically relevant interaction methods, including point annotations, bounding boxes, lasso tools, and scribbles, enhancing usability and clinical precision. In our evaluation involving 500 contrast-enhanced T1-weighted MRI scans from the BraTS 2025 Meningioma RT Segmentation Challenge, Interactive-MEN-RT demonstrated substantial improvement compared to other segmentation methods, achieving Dice similarity coefficients of up to 77.6% and Intersection over Union scores of 64.8%. These results emphasize the need for clinically tailored segmentation solutions in critical applications such as meningioma RT planning. The code is publicly available at: https://github.com/snuh-rad-aicon/Interactive-MEN-RT

Keywords: Magnetic Resonance Imaging · Meningioma · Tumor Segmentation · Interactive Medical Image Segmentation

1 Introduction

Meningiomas, the most common primary intracranial tumors, frequently require radiotherapy (RT) as a primary or adjuvant treatment modality. The success of RT is critically contingent upon the precise delivery of radiation to

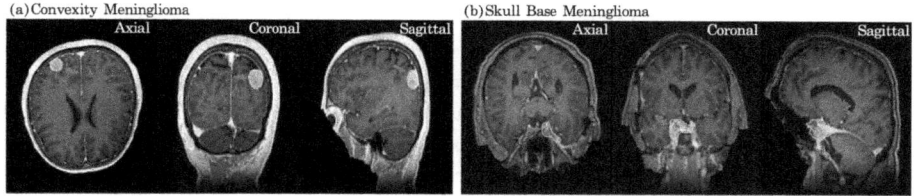

Fig. 1. Representative meningioma cases illustrating anatomical diversity. (a) Convexity meningioma with clear margins; (b) Skull base meningioma encasing structures

the target volume while minimizing exposure to adjacent healthy brain structures and organs at risk (OARs) [1]. Accurate delineation of the gross tumor volume (GTV) is therefore a cornerstone of effective RT planning. However, manual segmentation of meningiomas from medical imaging data-typically multiparametric Magnetic Resonance Imaging (MRI)-is a labor-intensive and technically demanding task. It is susceptible to substantial inter- and intra-observer variability, even among experienced radiation oncologists and neuroradiologists [2]. This inconsistency is primarily attributable to the anatomical diversity and intrinsic heterogeneity of meningiomas, including variations in tumor location, shape, and imaging features such as cystic degeneration, necrosis, or calcification.

The anatomical complexity and heterogeneity of meningiomas, particularly those at the skull base or near intricate neurovascular structures, pose significant challenges to accurate boundary delineation. Figure 1 illustrates representative examples of meningiomas in distinct intracranial locations, highlighting the marked variability in tumor morphology, size, and relationship to adjacent anatomical structures. As depicted, convexity meningiomas typically exhibit well-defined margins and limited involvement with critical neurovascular anatomy. In contrast, skull base or ventricular meningiomas often abut or encase major vessels and cranial nerves, rendering manual segmentation challenging even for experts. These examples underscore the need for advanced segmentation tools that can robustly accommodate both the anatomical diversity and internal complexity of meningiomas encountered in radiotherapy planning.

To address these limitations, automated segmentation approaches, particularly those leveraging deep learning, have garnered considerable attention. Convolutional Neural Networks (CNNs), especially U-Net-based architectures, have demonstrated state-of-the-art performance in diverse medical image segmentation tasks [3,4]. Nevertheless, the reliability of fully automated models remains suboptimal for cases involving heterogeneous tumor appearances or complex anatomical locations. Consequently, a persistent need for clinician oversight remains, especially in safety-critical applications like RT planning.

Interactive Medical Image Segmentation (IMIS) frameworks have emerged as a promising solution, integrating artificial intelligence (AI) with clinician expertise to facilitate more accurate and efficient tumor delineation [5]. The advent of foundation models such as the Segment Anything Model (SAM) [6,7] initially propelled progress in this domain, but early iterations were constrained by their two-dimensional scope. Recent advancements-including SAM-Med3D [8], MedSAM2 [9], and the nnInteractive framework [10]-have enabled interactive segmentation in volumetric medical images, fostering more effective clinicianAI collaboration. Despite these advances, many general-purpose IMIS tools treat the initial automated segmentation merely as a preliminary estimate, which may be insufficient for RT planning, where even minor errors can have significant dosimetric consequences.

In this study, we present Interactive-MEN-RT, a domain-specific and highly accurate IMIS tool tailored for meningioma delineation in the RT workflow. Built upon the robust nnU-Net V2 architecture [3,11] and enhanced with a bespoke interactive training module [10], our system is engineered to iteratively refine tumor segmentations in direct response to clinician input, thereby aligning the final contours with expert judgment. Our contributions are twofold: (1) We show that adapting IMIS models to the meningioma RT domain, using task-specific fine-tuning and clinician-guided iterative refinement, leads to significant improvements over general-purpose baselines like SAM-Med3D and nnInteractive in both segmentation accuracy and clinical usability. (2) We provide comprehensive validation demonstrating that, in the context of meningioma RT, specialized, domain-adapted models are essential to achieve the requisite accuracy and reliability for clinical use, with consistent gains observed across all forms of user interaction.

2 Methods

This section presents the methodology of the Interactive-MEN-RT system for accurate meningioma segmentation in radiotherapy planning. It describes the data preprocessing pipeline, the U-Netbased model architecture, the integration and simulation of interactive user prompts, and an ablation study evaluating the impact of transfer learning (TL). An overview of the entire pipeline is illustrated in Fig. 2.

2.1 Data Preprocessing and Preparation

For our experimental validation, we exclusively utilized the training set of the BraTS 2025 Meningioma RT Segmentation Challenge dataset [4]. This comprehensive dataset comprises 500 samples specifically curated for meningioma Gross Tumor Volume (GTV) segmentation in brain MRI. The dataset exclusively uses 3D contrast-enhanced T1-weighted (CE-T1w) images and preserves extracranial structures, with patient-identifying facial features removed via defacing techniques to ensure anatomical integrity. The entire dataset was partitioned into 400 samples for training and 100 for validation to facilitate robust model evaluation.

Our preprocessing pipeline adapted robust protocols from nnU-Net: (1) intensity normalization via Z-score standardization, (2) resampling to 1 mm isotropic

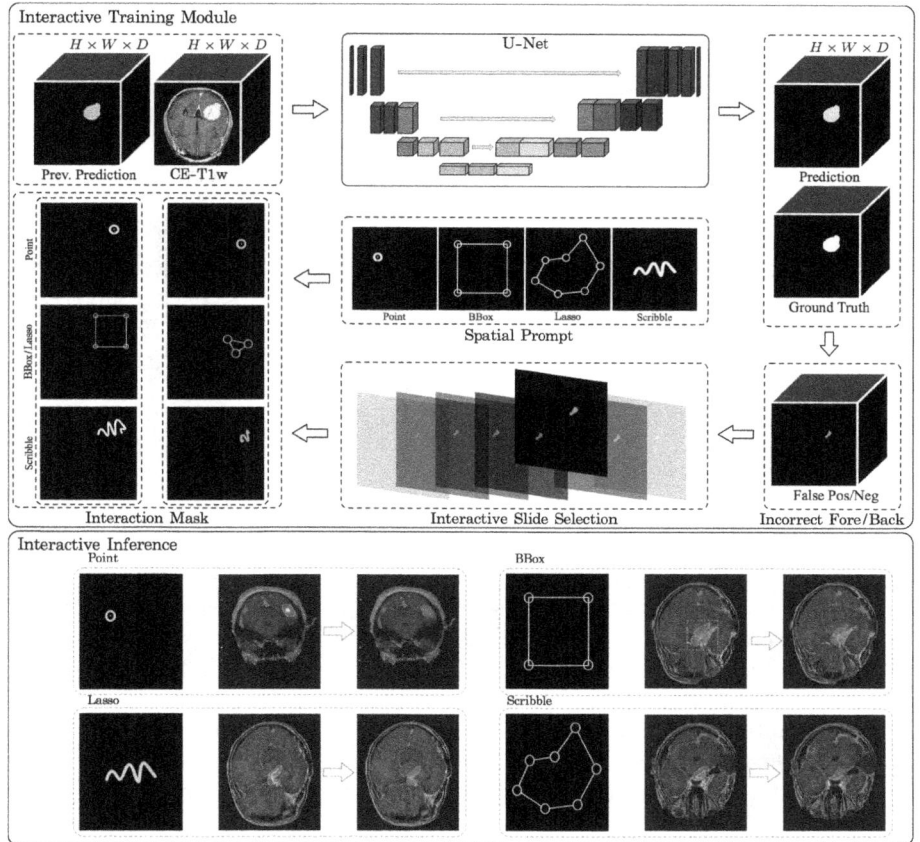

Fig. 2. An overview of the Interactive-MEN-RT for meningioma gross tumor volume segmentation with interactive prompts.

voxel spacing, (3) spatial cropping to eliminate excessive background while preserving anatomical context, and (4) data augmentation-including rotation, scaling, elastic deformation, and intensity variations-to enhance model robustness.

2.2 Network Architecture

The architecture of Interactive-MEN-RT is based on the principles of nnInteractive [10], a 3D promptable segmentation framework for volumetric medical image analysis. Consistent with nnInteractive's design, we employ an U-Net-based architecture over Transformer alternatives, building upon the nnU-Net framework [3,11,12] and using the Residual Encoder (ResEnc-L) configuration as its backbone. The architecture incorporates nnInteractive's sophisticated prompt-processing mechanisms, adapted for the specific requirements of meningioma segmentation to efficiently integrate user guidance. We leverage nnInteractive's multi-channel input design, which comprises three key components: (1) the original CE-T1w MRI data, (2) previous segmentation results, and (3) interactive guidance signals. Following nnInteractive's methodology, the network supports a comprehensive range of spatial prompt types, including points, bounding boxes, lasso selections, and scribbles [10,13]. Each prompt type is encoded in two separate input channels (positive and negative) within the interactive guidance signals.

We optimized the combination of loss functions for the binary task of differentiating tumor from background in a radiotherapy planning context. The final layer is configured for binary segmentation (tumor vs. background) with a sigmoid activation function, optimized using a combination of Dice loss and Cross-Entropy loss (DiceCELoss) to balance overlap accuracy with voxel-wise precision.

2.3 Interaction Prompts

User Interaction Types. Drawing from advancements in IMIS systems such as nnInteractive [10], our module supports intuitive 2D prompts on a standard axial view to guide 3D segmentation. The supported prompt types include points for correcting localized errors with positive foreground and negative background clicks; bounding boxes to define broad regions of interest; lasso selections as closed-loop contours for precise delineation of irregular tumor boundaries; and scribbles as free-form lines to indicate larger regions for inclusion or exclusion. These user interactions are converted into spatial maps that indicate areas of interest or correction, which are then combined with the original MRI to incorporate clinician expertise directly into the segmentation refinement process.

Training-Time Simulation of Prompts. During training, we simulated realistic user interactions to improve model robustness and clinical relevance. For each interaction type, including point, bounding box, scribble, and lasso prompts, we employed dedicated sampling strategies that emulate clinical correction workflows while introducing stochastic variations. For point prompts, we randomly sampled 12 positive points within the tumor, randomizing their location, size, and number. Bounding box prompts involved generating a single box on a tumor-containing slice with a random margin, often including jitter or size variation. Scribble prompts were created by connecting 28 random points within the tumor on a slice, with randomized order and added jitter or wavy effects to mimic freehand drawing. Finally, lasso prompts simulated polygons by sampling 412 jittered points along the tumor boundary on a selected slice, ensuring valid closed contours. Across all prompt types, slice selection was weighted by tumor area, and all randomizations captured the diversity of real-world clinical interactions.

2.4 Prompt Encoding

All user prompts (points, boxes, lassos, scribbles) are encoded as two additional channels: one for positive (foreground) and one for negative (background) interactions. Each channel matches the spatial dimensions of the input image and is normalized to the [0, 1] range. These prompt channels are concatenated with the original image to form the network input, allowing the model to flexibly incorporate user guidance.

2.5 Ablation Study: Transfer Learning

To evaluate the effect of transfer learning, we compared two model initialization strategies: (1) training the network from scratch with random initialization, and (2) using pre-trained nnInteractive weights for transfer learning. Both models were trained and evaluated under the same protocol and dataset settings.

2.6 Implementation Details

All models were implemented in PyTorch 1.13 and trained on a NVIDIA A6000 GPU with 48 GB of memory. Training was performed using 3D patches of size 128×128×128 and a batch size of 8. The optimizer was SGD with Nesterov momentum, an initial learning rate of 1e-2, and a polynomial decay schedule. Data augmentation included random rotations ($\pm 15°$), scaling (0.91.1×), elastic deformation, and intensity shifts ($\pm 10\%$). Consistent preprocessing and evaluation protocols were used for all experiments.

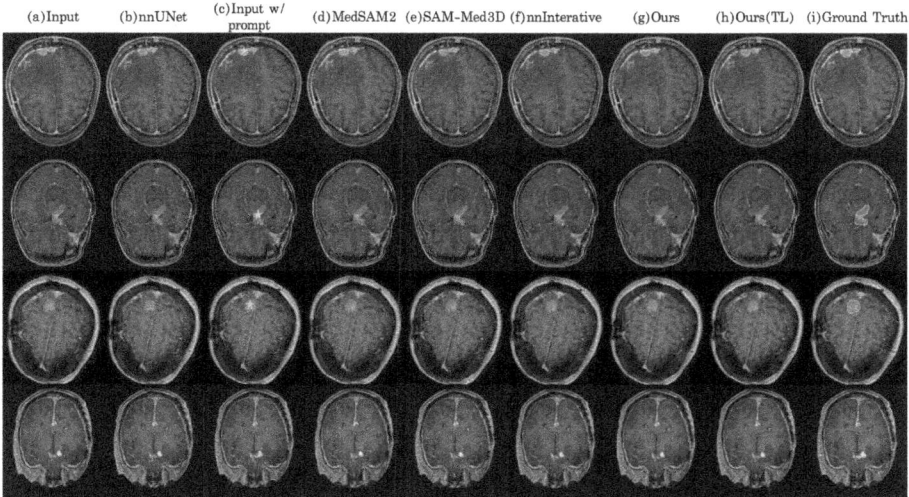

Fig. 3. Qualitative segmentation overlays for each method and the ground truth under point prompt interaction settings.

Table 1. Interactive segmentation performance by prompt type (mean ± SD).

Prompt	Method	DSC (%)	IoU (%)
None	nnUNet	65.5 ± 25.1	53.1 ± 23.9
Point	MedSAM2 [9]	10.1 ± 9.2	5.1 ± 5.0
	SAM-Med3D [8]	74.9 ± 21.9	60.4 ± 20.5
	nnInteractive	69.7 ± 22.2	57.2 ± 22.3
	Interactive-MEN-RT (scratch)	**75.5 ± 16.1**	**62.8 ± 17.2**
	Interactive-MEN-RT (TL)	72.8 ± 20.4	60.5 ± 20.6
BBox	MedSAM2 [9]	63.2 ± 21.5	49.1 ± 19.6
	nnInteractive	50.1 ± 16.4	35.0 ± 14.9
	Interactive-MEN-RT (scratch)	76.0 ± 13.0	62.9 ± 15.3
	Interactive-MEN-RT (TL)	**77.6 ± 11.2**	**64.6 ± 13.7**
Lasso	nnInteractive	61.8 ± 21.2	47.9 ± 20.9
	Interactive-MEN-RT (scratch)	**77.5 ± 13.0**	**64.8 ± 15.1**
	Interactive-MEN-RT (TL)	63.6 ± 30.2	52.6 ± 27.5
Scribble	nnInteractive	73.9 ± 17.7	61.2 ± 18.7
	Interactive-MEN-RT (scratch)	**76.2 ± 15.0**	**63.5 ± 16.5**
	Interactive-MEN-RT (TL)	69.7 ± 25.0	58.0 ± 23.7

2.7 Evaluation Metrics and Baselines

Segmentation accuracy was evaluated on the validation set using two standard metrics: the Dice Similarity Coefficient (DSC) and Intersection over Union (IoU). We compared the performance of Interactive-MEN-RT with several state-of-the-art interactive segmentation frameworks, including nnInteractive [10], SAM-Med3D [8], and MedSAM2 [9], as well as an ablation variant that distinguished between training from scratch and transfer learning. To compare with an automatic segmentation model, we additionally included nnUNet [14] as a strong baseline due to its proven generalization capability across diverse medical segmentation tasks. To evaluate the interactive methods, user interactions were simulated by sampling prompts from ground truth tumor regions on a per-case basis.

3 Results

3.1 Qualitative Assessment by Prompt Type

As shown in the interactive inference results in Fig. 2, Interactive-MEN-RT consistently produces accurate and robust segmentations across diverse prompt types. The visual examples demonstrate that Interactive-MEN-RT consistently achieves high-quality segmentations regardless of the prompt strategy employed. Interactive-MEN-RT adeptly handles various input modalities, from point prompts for initial delineations to bounding box prompts for refining complex regions. This adaptability is crucial for real-world clinical scenarios where diverse tumor characteristics and user preferences necessitate flexible interaction.

Fig. 4. Qualitative 3D segmentations for each method and the ground truth.

3.2 Interactive Segmentation Performance

Table 1 presents a comparative performance analysis of Interactive-MEN-RT against leading interactive segmentation models, including SAM-Med3D [8], MedSAM2 [9], and nnInteractive [10]. The analysis evaluates segmentation accuracy (DSC and IoU) across various prompt types: points, bounding boxes, lassos, and scribbles. This comparison also serves as an ablation study, highlighting the impact of transfer learning within the Interactive-MEN-RT framework.

Interactive-MEN-RT consistently demonstrated performance superior or comparable to the baseline methods across all prompt types. With point-based prompts, the scratch-trained Interactive-MEN-RT model yielded the highest DSC (75.5% ± 16.1), outperforming all baselines, while its transfer learning (TL) variant achieved a leading IoU score (60.5% ± 20.6). For bounding box interactions, the TL model achieved the highest DSC (77.6% ± 11.2) and IoU (64.6% ± 13.7), with the scratch-trained model also showing strong performance. In the case of lasso prompts, the scratch-trained model again led in both DSC (77.5% ± 13.0) and IoU (64.8% ± 15.1), substantially outperforming both nnInteractive and the TL variant. Similarly, for scribble-based corrections, the scratch model obtained the highest DSC (76.2% ± 15.0) and IoU (63.5% ± 16.5).

3.3 Comparative Qualitative Analysis of Segmentation Methods

Figure 3 presents a qualitative comparison of segmentation results obtained under a single point prompt on representative meningioma cases. The figure facilitates a direct visual comparison of outputs from MedSAM2, SAM-Med3D, nnInteractive, Interactive-MEN-RT (trained from scratch), and Interactive-MEN-RT (with transfer learning) against the ground truth annotations. Each row corresponds to a distinct meningioma case, strategically chosen to exemplify the diversity of tumor locations relevant to RT planning: convexity, skull base, falx/parasagittal, and ventricular regions. This presentation of qualitative results across varied clinical scenarios effectively demonstrates the robustness and adaptability of Interactive-MEN-RT.

As shown in Fig. 4, volumetric renderings of predicted segmentations illustrate the spatial coherence and completeness of Interactive-MEN-RT compared to baseline models, offering a more comprehensive perspective on tumor coverage that may be particularly helpful for irregular or deeply situated lesions in anatomically complex regions such as the skull base and ventricles.

4 Discussion

This study demonstrates the clinical and technical value of Interactive-MEN-RT, a specialized interactive segmentation system designed for meningioma radiotherapy planning. It enables clinicians to delineate target volumes with minimal input, reducing inter-observer variability and streamlining RT workflows. Its robustness across diverse prompt types-including points, bounding boxes, lassos, and scribbles-ensures flexibility in accommodating varying user preferences and clinical scenarios. Importantly, the ability to achieve high accuracy with minimal input makes it well-suited for time-sensitive clinical workflows.

Our comparative analysis demonstrates that Interactive-MEN-RT consistently achieves superior or competitive performance compared to established baseline models, such as SAM-Med3D [8], MedSAM2 [9], and nnInteractive [10]. The most substantial performance improvements were observed with point and lasso prompts, emphasizing the effectiveness of our prompt encoding and training methodologies. These results highlight the value of domain-specific models in safety-critical tasks where general-purpose tools may fall short.

Models trained from scratch on task-specific data often surpassed transfer learning, especially with fine-grained prompts. Although nnInteractive generalizes well, it underperforms in capturing lesion-specific details compared to scratch-trained models. This limitation likely arises from the model's optimization towards particular prompt types or anatomical contexts, restricting its adaptability during fine-tuning. Transfer learning remains useful when data or compute is limited, especially for simple prompts like boxes. A hybrid strategy combining transfer learning and domain-specific training is recommended for precision-critical applications.

This study has several limitations. First, user interactions were simulated rather than obtained directly from clinicians, potentially limiting the representation of the diverse and complex nature of real-world clinical usage. Second, the evaluation was limited to a single dataset [4], raising concerns about generalizability. Finally, prospective clinical studies and comprehensive user experience assessments are needed to confirm the system's practical utility and usability in clinical settings.

5 Conclusion

In conclusion, Interactive-MEN-RT presents a promising interactive segmentation solution tailored for meningioma radiotherapy planning, achieving strong performance across various prompt types. Its ability to deliver accurate segmentation with minimal input, combined with favorable comparisons against established baselines, suggests strong potential to enhance efficiency, consistency, and safety in clinical radiotherapy workflows. These findings underscore the value of utilizing specialized, disease-specific models in safety-critical medical settings, as they may better address the limitations of general-purpose frameworks and align more closely with clinical practice demands.

Disclosure of Interests. The authors declare that they have no competing interests.

References

1. Goldsmith, C., et al.: ESTRO ACROP guideline for target delineation of skull base meningiomas. Radiother. Oncol. **128**(3), 433–439 (2018)
2. Mabray, M.C., et al.: Variability of meningioma volumetric measurements and interobserver agreement. Neuroradiology **57**(10), 1003–1010 (2015)
3. Isensee, F., et al.: nnU-Net: a self-configuring method for deep learning-based biomedical image segmentation. Nat. Methods **18**(2), 203–211 (2021)
4. LaBella, D et al.: Brain Tumor Segmentation (BraTS) Challenge 2024: Meningioma Radiotherapy Planning Automated Segmentation. arXiv preprint (2024)
5. Cheng, J., et al.: Interactive medical image segmentation: A benchmark dataset and baseline. arXiv preprint arXiv:2411.12814 (2024)
6. Kirillov, A., et al.: Segment Anything. arXiv preprint (2023)
7. Ravi, N., et al.: Sam 2: Segment anything in images and videos. arXiv preprint arXiv:2408.00714 (2024)
8. Wang, H., et al.: SAM-Med3D: towards General-purpose Segmentation Models for Volumetric Medical Images. arXiv preprint (2023)
9. Ma, J., et al.: MedSAM2: Segment Anything in 3D Medical Images and Videos. arXiv preprint (2025)
10. Isensee, F., et al.: nnInteractive: Redefining 3D Promptable Segmentation. arXiv preprint (2025)
11. Isensee, F., Jaeger, P.F., Kohl, S.A.A., Petersen, J., Maier-Hein, K.H.: nnU-Net: Self-adapting Framework for U-Net Based Medical Image Segmentation. Nature Methods (2021), nnU-Net V2 details, see GitHub repository
12. Vaswani, A., et al.: Attention is all you need. Adv. Neural Inform. Process. Syst. **30** (2017)
13. Wong, H.E., Rakic, M., Guttag, J., Dalca, A.V.: Scribbleprompt: fast and flexible interactive segmentation for any biomedical image. In: European Conference on Computer Vision. pp. 207–229. Springer (2024). https://doi.org/10.1007/978-3-031-73661-2_12
14. Isensee, F., Jaeger, P.F., Kohl, S.A., Petersen, J., Maier-Hein, K.H.: nnu-net: a self-configuring method for deep learning-based biomedical image segmentation. Nat. Methods **18**(2), 203–211 (2021)

AI-Driven Multimodal TMJ Patient Modeling: From Unstructured Notes to Precision Treatment

Alban Gaydamour[1,2(✉)], Enzo Tulissi[1,2], Claudia Mattos[3], Rodrigo Teixeira[4], Maxwell Shin[1], Adam Hershey[1], Anabelle Kwon[1], Felicia Miranda[4], Marcela Gurgel[5], Selene Barone[6], Aron Aliaga[1], Marilia Yatabe[1], Paulo Zupelari[1], Marina Zupelari[1], David Hanauer[1], Nina Hsu[1], Steve Pieper[7], Eduardo Caleme[8], Jonas Bianchi[9], Joao Goncalves[10], Daniela Goncalves[10], Lawrence Wolford[11], Antonio Ruellas[12], Juan Prieto[13], Tengfei Li[13], Hongtu Zhu[13], Runpeng Dai[13], Martin Styner[13], Najla Al Turkestani[14], Alexandre F. DaSilva[1], and Lucia Cevidanes[1]

[1] University of Michigan, Ann Arbor, MI, USA
[2] CPE Lyon, Lyon, France
albangay@umich.edu
[3] Fluminense Federal University, Rio de Janeiro, Brazil
[4] University of São Paulo, São Paulo, Brazil
[5] Federal University of Ceara, Fortaleza, Brazil
[6] Magna Græcia University, Catanzaro, Italy
[7] Isomics Inc., Cambridge, MA, USA
[8] Positive University, Curitiba, Brazil
[9] University of the Pacific, San Francisco, CA, USA
[10] State University of São Paulo, São Paulo, Brazil
[11] Baylor University Medical Center, Dallas, TX, USA
[12] Federal University of Rio de Janeiro, Rio de Janeiro, Brazil
[13] University of North Carolina, Chapel Hill, NC, USA
[14] King AbdulAziz University, Jeddah, Saudi Arabia

Abstract. Temporomandibular degenerative joint disease (TM DJD) is a multifactorial condition with complex clinical presentations. This study presents a multimodal framework centered on structured summarization of clinical text, supported by imaging information from automatically registered MRI and CBCT scans. Two large language models, BART and DeepSeek-R1, were fine-tuned on 1,813 annotated text segments from 500 TM DJD patient records to extract 56 clinical indicators, including pain severity, jaw function, imaging findings, and sleep disturbances. The models converted narrative notes into structured data fields for use in clinical dashboards enabling patient-specific and population-level analyses. BART outperformed DeepSeek in clinical field extraction accuracy, precision, and recall, despite DeepSeek achieving slightly higher ROUGE metrics based on word-level overlap. A parallel automated MRI-to-CBCT registration pipeline achieved submillimeter accuracy and a 98.75% success rate. This work extracted clinically meaningful pain comorbidities and radiological findings from unstructured clinical narratives, enabling actionable insights for musculoskeletal precision care. The future integration of structured clinical data and multimodal image analyses may enable holistic, personalized patient models.

Keywords: Temporomandibular degenerative joint disease · Multimodal imaging · Large Language Models

1 Introduction

Temporomandibular disorders (TMD) are the second most prevalent musculoskeletal condition after chronic low back pain, affecting 5–12% of the population and costing an estimated $4 billion annually [1]. Around 80% of patients exhibit clinical signs such as disc displacement and joint pain often progressing to temporomandibular degenerative joint disease (TM DJD) [1–3]. Comorbidities like headaches and sleep disturbances are common and significantly influence disease progression and treatment outcomes [4–6]. While MRI and CBCT are essential for visualizing joint anatomy, they do not capture a patient's symptom burden or comorbidity profile, factors critical for personalized treatment. Clinical decision-making requires understanding both structural changes and functional impact, such as sleep disruption, jaw locking, or radiating pain. These details, often documented only in free-text notes, were captured at scale using MedEx, a tool developed with two fine-tuned large language models (LLMs): Bidirectional Auto-Regressive Transformer (BART) [10] and DeepSeek-R1. Both models were downloaded and used offline to ensure patient data privacy, with no clinical information transmitted over the internet. Trained on 500 TMD notes processed into 1813 annotated segments, both models extract structured data from unstructured narratives [7–9], mapping each note to 56 predefined clinical fields (e.g., "maximum opening: 38 mm"; "disc displacement: with or without reduction"). This work provides a structured textual layer containing disease-related pain comorbidities and imaging findings, along with MRI-to-CBCT registered images [11]. These datasets support comprehensive TMJ evaluation and are aligned with initiatives such as the TMD IMPACT Consortium [12] and the NIH HEAL Initiative [13].

2 Methods

2.1 Study Sample

This study focuses on the structured summarization of clinical notes for patients with TM DJD. A retrospective dataset of de-identified clinical notes was compiled from 500 patients treated at the Baylor University Department of Oral and Maxillofacial Surgery. The dataset includes initial clinical examination notes and radiology reports for CBCTs and MRIs, capturing pain-related comorbidities and diagnostic impressions at baseline diagnosis. Patients were included only if MRI and CBCT imaging had been performed within a one-year interval to ensure temporal alignment between modalities. The imaging pipeline incorporated a previously validated MRI-to-CBCT registration method [11]. This method aligns MRI with CBCT scans using rigid mutual informationbased registration, achieving submillimeter translation errors and rotation differences under 3 degrees. Follow-up visits and patients under 12 years of age were excluded. This study was approved as exempt by the University of Michigan Institutional Review Board (HUM00239207).

2.2 Data Preprocessing and Annotation

Clinical notes (PDF/DOCX) were converted to plain text via a custom Python script. To meet large language model input constraints, each document was segmented into discrete, non-overlapping text segments, with a maximum length of 1024 tokens. Text segmentation to 1024 tokens was required due to input size constraints of the BART model. While DeepSeek supports larger context windows, identical chunk sizes were used across models to ensure fair comparison and maintain consistency. Segmentation was designed to preserve semantic integrity, aggregating full paragraphs and breaking only at sentence boundaries when necessary. Annotation guidelines were defined through a clinician-led calibration process to ensure consistent labeling. These guidelines were then applied to annotate all 1813 segments using a predefined set of 56 descriptors related to TM DJD, including pain, sleep disturbances, hearing loss, and jaw function. Only terms with explicit textual evidence were included in the final dataset, which was then used to train and evaluate multiple summarization models. This textual layer complements the anatomical context captured through MRI-CBCT registration, without requiring additional image-based annotation (Fig. 1).

2.3 Model Architectures and Training Details

Two large language models were selected to evaluate clinical summarization performance: BART-large-CNN [14], a summarization-specific encoder–decoder model, and DeepSeek-R1-Distill-Qwen-1.5B, a distilled general-purpose language model (Figs. 2 and 3). It is important to note that although BART is a summarization-specific model, the training objective was not to generate shorter narrative summaries, but rather to produce structured outputs (key-value pairs). These models were chosen for their wide adoption and architectural diversity, enabling comparison between task-specific and general-purpose modeling approaches. BART was obtained from the Hugging Face Transformers library and fine-tuned using its default summarization configuration. DeepSeek was fine-tuned using Low-Rank Adaptation (LoRA) and 4-bit quantization. LoRA adapters were applied to key attention and MLP components to enable adaptation of structural and content representations for summarization. Both models were trained on the same dataset of annotated clinical text segments. Training was conducted using cosine learning rate scheduling, gradient accumulation over 4 steps, and without any prompt engineering or instruction tuning.

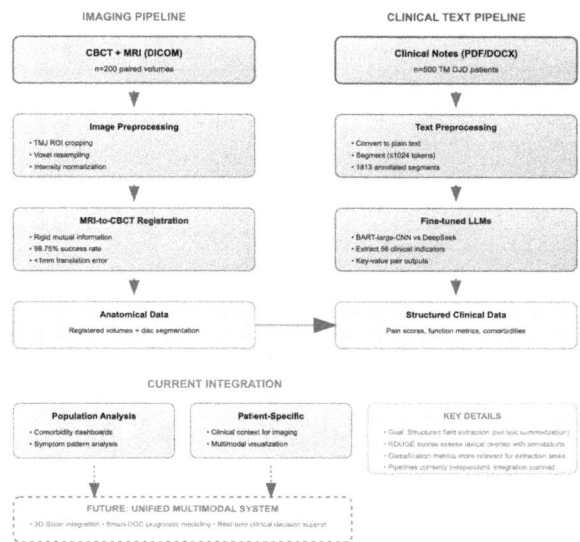

Fig. 1. Overview of the multimodal TMJ analysis workflow. The imaging pipeline (left) processes CBCT and MRI volumes through preprocessing and registration, enabling disc segmentation; over 200 patients have been processed to date. The clinical text pipeline (right) extracts 56 structured diagnostic fields from unstructured notes using fine-tuned LLMs, applied to 500 patients. Unlike traditional summarization, the models perform structured field extraction. Current outputs support both population-level comorbidity analysis and patient-specific interpretation. ROUGE scores assess lexical similarity, while classification metrics reflect clinical extraction accuracy. Future work will unify these tools within 3D Slicer for real-time diagnostic support.

Model outputs were evaluated using Recall-Oriented Understudy for Gisting Evaluation (ROUGE) metrics to assess summarization quality [15].

2.4 Training and Validation Protocol

The dataset was divided using an 80:10:10 split (80% of samples for training, 10% for validation, and 10% for testing). A 5-fold cross-validation protocol was employed to mitigate overfitting and assess generalizability. Both BART and DeepSeek models were evaluated across all folds under identical conditions. Training was monitored via ROUGE scores, and the model with the best ROUGE-1 score across folds was designated as the final summarization model.

2.5 Performance Evaluation

Model performance was evaluated using two categories of metrics: ROUGE-based lexical similarity and classification metrics for clinical field extraction.

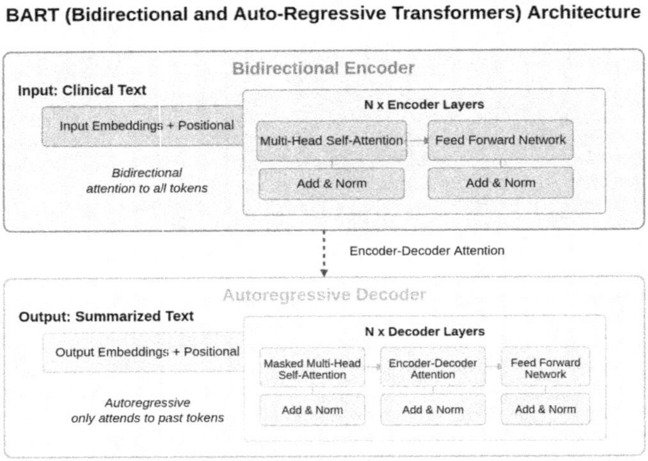

46   A. Gaydamour et al.

Fig. 2. Architecture of the BART LLM used for fine-tuning.

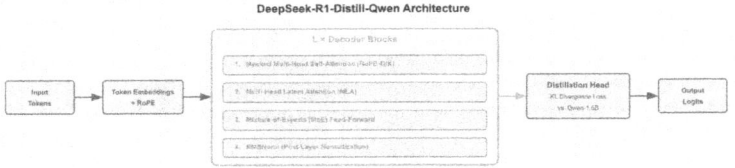

Fig. 3. Architecture of the DeepSeek LLM used for fine-tuning.

ROUGE-1, ROUGE-2, ROUGE-L, and ROUGE-Lsum were used to assess overlap between predicted and reference summaries. Additionally, accuracy, precision, recall, and F1 scores were computed for ten representative clinical categories, including headache intensity, jaw function, and disability rating. These values were derived using structured post-processed outputs and corresponding clinician annotations. Particular attention was given to comorbidity indicators with variable expression patterns, which tend to challenge LLMs in medical documentation. Both the BART and DeepSeek models were evaluated under this framework. The extracted summaries were formatted as keyvalue pairs (e.g., `maximum opening: 48mm`, `daily pain: 6`), enabling alignment with image-based dashboards and structured visualization tools.

3 Results

The MRI-to-CBCT registration framework previously validated on 70 paired volumes achieved a 98.75% success rate, with mean translation error below 1 mm and rotational deviation under 3° [11]. Figure 4 and Table 1 demonstrate this high-accuracy pipeline that provides a reliable anatomical foundation for multimodal TMJ analysis.

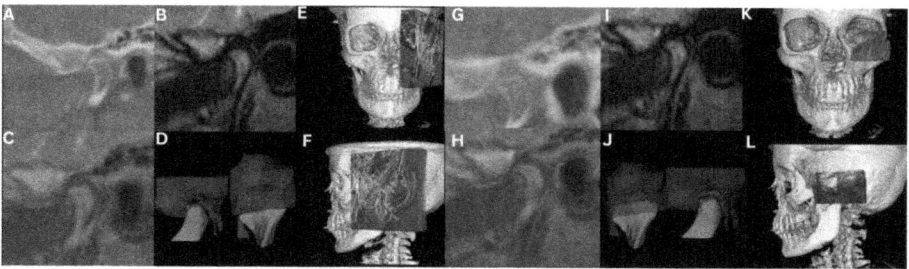

Fig. 4. MRI-CBCT registration comparison: A, G) Fixed CBCT, B, I) MRI manual/automated registration, C, H) MRI-CBCT overlays, D, J) TMJ segmentation (mandible: red, cranial base: blue, disc: green), E, K) Frontal views, F, L) Lateral views. (Color figure online)

Tables 2 and 3 report ROUGE summarization scores under 5-fold cross-validation for the BART and DeepSeek models, respectively. DeepSeek slightly outperformed BART in ROUGE scores (e.g., ROUGE-L +1.5%), indicating a minor advantage in surface-level fluency and lexical overlap. However, ROUGE metrics primarily assess general summarization quality and do not directly reflect field-level extraction accuracy required for clinical deployment.

Tables 4 and 5 present the classification performance of both models on 10 comorbidity-related clinical fields. Despite lower ROUGE scores, BART achieved substantially higher accuracy, precision, recall, and F1 scores across most categories. BART's F1 scores exceeded 70% in structured fields such as patient age, airway obstruction, and arthritis location. In contrast, DeepSeek performance varied more widely and was notably lower on low-prevalence items such as tinnitus and disability rating.

To illustrate the practical value of the extracted information, the manually curated clinical summaries for the full 500-patient cohort were compiled and analyzed. Structured key-value pairs—such as patient age, headache intensity, and headache location—were aggregated across the dataset. Descriptive statistics including prevalence, mean scores, and standard deviations were computed for relevant fields. This analysis revealed consistent TM DJD comorbidity patterns, including high prevalence of sleep disturbances, lateralized jaw pain, and variation in headache intensity. Figure 5 presents a visual dashboard generated from the manually extracted summaries, offering insight into population-level symptom trends and TMJ function characteristics.

Table 1. Differences in six Degrees of Freedom between clinician and Elastix Image Registration.

ROTATION (°)			
	Pitch	Roll	Yaw
Mean difference (SD)	0.49 (2.28)	−0.62 (2.21)	−0.99 (3.84)
Mean absolute difference (SD)	1.53 (1.75)	1.69 (1.54)	2.70 (2.89)
Minimum absolute difference	0.00	0.02	0.03
Maximum absolute difference	8.72	6.79	16.60
75th percentile of absolute difference	1.93	2.56	3.04
90th percentile of absolute difference	3.51	3.32	5.10
TRANSLATION (mm)			
	LR	AP	SI
Mean difference (SD)	0.3 (1.86)	0.89 (0.94)	−0.20 (0.63)
Mean absolute difference (SD)	0.92 (1.64)	0.98 (0.85)	0.50 (0.43)
Minimum absolute difference	0.01	0.01	0.00
Maximum absolute difference	13.10	3.52	2.57
75th percentile of absolute difference	1.16	1.48	0.72
90th percentile of absolute difference	1.55	2.12	0.95

Table 2. Summarization performance (ROUGE scores) of fine-tuned BART across 5-fold cross-validation.

	ROUGE-1	ROUGE-2	ROUGE-L	ROUGE-Lsum
Fold 1	83.68	71.99	83.50	83.49
Fold 2	83.48	73.40	83.11	83.14
Fold 3	84.93	74.23	84.38	84.57
Fold 4	85.50	74.73	85.11	85.21
Fold 5	85.47	74.64	84.98	85.01
Average	84.61	73.80	84.22	84.29

Table 3. Summarization performance (ROUGE scores) of fine-tuned DeepSeek across 5-fold cross-validation.

	ROUGE-1	ROUGE-2	ROUGE-L	ROUGE-Lsum
Fold 1	86.55	86.49	86.53	86.54
Fold 2	84.90	84.79	84.86	84.82
Fold 3	86.08	86.09	86.10	86.11
Fold 4	85.96	85.91	85.95	85.92
Fold 5	85.21	85.19	85.17	85.21
Average	85.74	85.70	85.72	85.72

Table 4. Fine-tuned BART Model Performance for Best Fold

	Accuracy (%)	Precision (%)	Recall (%)	F1 score (%)
Patient Age	78.6	88.0	88.0	88.0
Airway Obstruction	65.3	78.1	80.0	79.0
Arthritis Location	62.5	72.3	81.4	76.9
Headache Intensity	54.4	77.5	64.6	70.4
Fibromyalgia Present	50.0	100.0	50.0	66.7

Table 5. Fine-tuned DeepSeek Model Performance for Best Fold

	Accuracy (%)	Precision (%)	Recall (%)	F1 score (%)
Jaw Function	51.1	84.4	61.4	71.1
Arthritis Location	49.2	64.0	68.1	66.0
Earache	35.7	72.3	60.6	52.6
Airway Obstruction	29.9	40.0	54.1	46.0
Muscle Tenderness	25.4	32.6	53.6	40.6

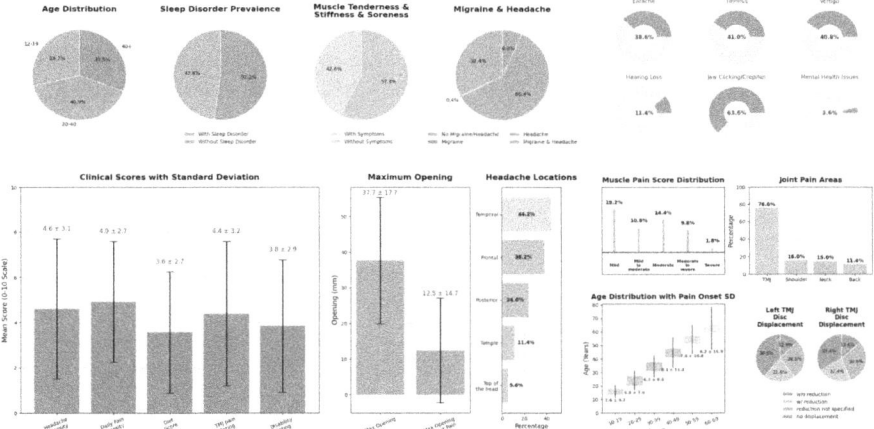

Fig. 5. Population-level dashboard of TMJ comorbidities (e.g., pain severity, jaw function) extracted from 500 clinical notes.

4 Discussion

This study presents a framework that combines an automated MRI-to-CBCT registration pipeline with structured clinical note summarization to support comprehensive assessment of TM DJD. Building on prior work [11], the imaging pipeline now automates previously manual steps such as initial approximation and TMJ cropping, thereby reducing inter-observer variability and enhancing reproducibility. Segmentations were first generated using an automated AI-based approach, and then manually reviewed and corrected by trained clinicians to ensure high-quality anatomical accuracy prior to registration. The workflow is implemented as an open-source 3D Slicer module, facilitating both clinical and research use. Validated on paired MRI-CBCT volumes, the registration achieved a 98.75% success rate with submillimeter translation and sub3° rotational error, yielding anatomically coherent 3D representations of the TMJ [11].

This work focuses on the automated extraction of structured information from clinical and imaging notes. A dataset of 500 de-identified notes from TM DJD patients was used to fine-tune two LLMs: BART and DeepSeek-R1. While DeepSeek achieved marginally higher ROUGE scores, BART consistently outperformed on field-level metrics including accuracy, precision, and recall. These findings are consistent with prior research showing that BART can outperform general-purpose models by over 15% in ROUGE-L for EHR summarization [10], and that smaller, domain-adapted models provide more reliable performance in precision-sensitive clinical NLP tasks [8].

A key strength of this framework lies in MedEx's ability to navigate diverse documentation styles. Given the absence of standardized questionnaires across TMD centers, clinical heterogeneity often hinders comparability. MedEx's structured outputs help normalize unstructured documentation and enable aggregation of pain, function, and sleep metrics across a 500-patient dataset. This structured approach enables direct correlation between functional limitations and anatomical findings. Dashboards constructed from these summaries highlighted trends in headache intensity, lateralized joint pain, and functional limitation [16], reflecting the model's robustness in capturing fragmented or variable text. The structured summaries also support patient-specific visualization and analysis. Each individual's comorbidities, such as joint arthritis, headache location, and airway obstruction, are aligned with their corresponding MRI-CBCT imaging findings, enabling a unified, subject-level diagnostic view.

Several areas of improvement remain. Comparison with a naive baseline using EMERSE, a text-mining tool for clinical notes, is ongoing to quantify MedEx's added value. Additionally, although BART and DeepSeek were selected for task-specific and general-purpose comparison, future work will assess zero-shot LLMs like GPT-4 and Gemini using PHI-safe platforms. These models may enable few-shot prompting or direct deployment in resource-constrained environments. Planned development will align the structured outputs from LLMs with 3D Slicer dashboards that integrate MRI-CBCT visualizations, where each patient's extracted comorbidities will be displayed alongside their anatomical imaging data, enabling personalized, multimodal planning.

Another challenge lies in the model's sensitivity to variations in note structure and terminology, which can limit generalizability. Errors may stem from underrepresented terms or occasional hallucinations [17]. Planned data augmentation includes structural and lexical modifications-such as synonym replacement, noise injection, and domain-specific paraphrasing-drawing from denoising pretraining strategies [18]. These techniques can enhance robustness while preserving clinical validity. Future work will explore longer-context models (e.g., DeepSeek's full context window, GPT-4) to evaluate whether document-level coherence improves field extraction, especially for cross-sentence inferences.

Although ROUGE and classification scores offer valuable benchmarks for performance, clinical deployment will require human-in-the-loop evaluation and broader generalization across patient populations. By aligning imaging data with structured summaries extracted from clinical notes, this work establishes a scalable foundation for TMJ analysis that links radiologic context with diagnostic information captured in text, enhancing the utility of both data sources for clinical decision-making. Future directions include automated segmentation of the articular disc from MRI, comparisons with rule-based systems such as EMERSE, and evaluation of zero-shot LLMs for clinical deployment. Outputs such as CSV files and dashboards facilitate compatibility with clinical workflows and integration into open-source visualization tools [19].

5 Conclusion

This study introduces a multimodal framework for TMJ assessment that combines automated MRI-to-CBCT registration with structured clinical summarization using fine-tuned language models. Summarization models extract comorbidity and imaging related indicators from clinical notes, with BART outperforming DeepSeek in structured output accuracy. The resulting structured datasets and visual dashboards reveal clinically relevant patterns in pain, function, and sleep disturbances, supporting population-level analysis and providing integrated visualization of patient-specific diagnoses.

Acknowledgments. This work was funded by NIH, grant number R01-DE024450.

Disclosure of Interests. The authors have no competing interests to declare that are relevant to the content of this article.

References

1. Gatchel, R.J., Stowell, A.W., Wildenstein, L., Riggs, R., Ellis, E.: Efficacy of an early intervention for patients with acute temporomandibular disorder-related pain: a one-year outcome study. J. Am. Dent. Assoc. **137**(3), 339–347 (2006)
2. Plesh, O., Sinisi, S.E., Crawford, P.B., Gansky, S.A.: Diagnoses based on the research diagnostic criteria for temporomandibular disorders in a biracial population of young women. J. Orofac. Pain **19**, 65–75 (2005)
3. Manfredini, D., Segu, M., Bertacci, A., Binotti, G., Bosco, M.: Diagnosis of temporomandibular disorders according to RDC/TMD axis I findings, a multicenter Italian study. Minerva Stomatol. **53**, 429–438 (2004)
4. Pihl, K., Roos, E.M., Taylor, R.S., Grønne, D.T., Skou, S.T.: Associations between comorbidities and immediate and one-year outcomes following supervised exercise therapy and patient education - A cohort study of 24,513 individuals with knee or hip osteoarthritis. Osteoarthr. Cartil. **29**(1), 39–49 (2021)

5. Al Turkestani, N., Li, T., Bianchi, J., et al.: A comprehensive patient-specific prediction model for temporomandibular joint osteoarthritis progression. Proc. Natl. Acad. Sci. U.S.A. **121**(8), e2306132121 (2024)
6. Scattergood, S.D., Cheng, V., Wylde, V., Blom, A.W., Whitehouse, M.R., Lenguerrand, E.: Influence of pre-operative co-morbidities on pain and function outcomes at 1 year after primary total knee arthroplasty. Knee **54**, 263–274 (2025)
7. Schiffman, E., Ohrbach, R., Truelove, E., et al.: Diagnostic criteria for temporomandibular disorders (DC/TMD) for clinical and research applications: recommendations of the international RDC/TMD consortium network and orofacial pain special interest group. J. Oral Facial Pain Headache **28**(1), 6–27 (2014)
8. Ge, J., Li, M., Delk, M.B., Lai, J.C.: A comparison of a large language model vs manual chart review for the extraction of data elements from the electronic health record. Gastroenterology **166**(4), 707–709 (2024)
9. Hsu, E., Roberts, K.: LLM-IE: a python package for biomedical generative information extraction with large language models. JAMIA Open **8**(2), ooaf012 (2025)
10. Zhou, F., Qin, B., Lan, G., Ye, Z.: News text generation method integrating pointer-generator network with bidirectional auto-regressive transformer. In: 2023 2nd International Conference on Artificial Intelligence and Intelligent Information Processing (AIIIP). IEEE, pp. 114–118 (2023)
11. Leroux, G., et al.: Novel CBCT-MRI registration approach for enhanced analysis of temporomandibular degenerative joint disease. In: Drechsler, K., Oyarzun Laura, C., Freiman, M., Chen, Y., Wesarg, S., Erdt, M. (eds.) Clinical Image-Based Procedures. CLIP 2024. LNCS, vol. 15196, pp. 63–72. Springer, Cham (2024). https://doi.org/10.1007/978-3-031-73083-2_7
12. National Institute of Dental and Craniofacial Research: TMD Collaborative for Improving Patient-Centered Translational Research (TMD IMPACT). https://www.nidcr.nih.gov/grants-funding/research-funded-by-nidcr-extramural/tmd-impact. Accessed 18 Mar 2025
13. NIH Heal Initiative. https://heal.nih.gov/. Accessed 18 Mar 2025
14. Lewis, M., Liu, Y., Goyal, N., et al.: BART: denoising sequence-to-sequence pretraining for natural language generation, translation, and comprehension. In: Proceedings of the 58th Annual Meeting of the Association for Computational Linguistics, pp. 7871–7880 (2020)
15. Lin, C.Y.: ROUGE: a package for automatic evaluation of summaries. In: Text Summarization Branches Out, pp. 74–81 (2004)
16. Dowding, D., Randell, R., Gardner, P., et al.: Dashboards for improving patient care: review of the literature. Int. J. Med. Inform. **84**(2), 87–100 (2015)
17. Huang, L., Yu, W., Ma, W., et al.: A survey on hallucination in large language models: principles, taxonomy, challenges, and open questions. ACM Trans. Inf. Syst. **43**(2), 1–55 (2025)
18. Chen, X., Long, G., et al.: Improving the robustness of summarization systems with dual augmentation. In: Proceedings of the 61st Annual Meeting of the Association for Computational Linguistics (Volume 1: Long Papers), pp. 6846–6857 (2023)
19. Beam, A.L., Kohane, I.S.: Big data and machine learning in health care. JAMA **319**(13), 1317–1318 (2018)

OrificeNet: Automatic Concealed Orifice Detection from Microscope Imagery with CBCT-Guided Refinement

Kefan Zhou[1], Yufei Chen[1]([✉]), Wei Liu[1], Qiyun Shen[2], and Qi Zhang[2]

[1] School of Computer Science and Technology,
Tongji University, Shanghai, China
yufeichen@tongji.edu.cn

[2] Department of Endodontics, School and Hospital of Stomatology, Tongji University, Shanghai Engineering Research Center of Tooth Restoration and Regeneration, Shanghai, China

Abstract. Accurate localization of all root canal orifices is critical for the success of root canal treatment. However, concealed orifices, particularly the second mesiobuccal canal (MB2), are frequently missed due to ambiguous visual cues, leading to high rates of treatment failure. To address this challenge, we propose a novel computer-aided detection framework, OrificeNet, the first to perform orifice detection directly from intraoperative microscope-view RGB images. Our framework formulates this as a segmentation task, employing an encoder-decoder network with a multi-scale strategy and a hierarchical cascaded decoder to effectively identify orifices. Furthermore, to simulate the real clinical workflow, we introduce a CBCT-guided post-processing step that leverages pre-operative 3D data to refine the 2D prediction via an affine transformation, accurately locating and completing concealed orifices. Extensive experiments on a clinically collected dataset demonstrate that our proposed method significantly achieve better performance. Our work presents an effective and clinically-translatable solution to reduce the risk of missed canals, enhancing the success rate of root canal treatments.

Keywords: Canal orifices detection · Deep learning

1 Introduction

Root canal treatment aims to eliminate infection and preserve the natural tooth by thoroughly cleaning and sealing the root canal system. A critical prerequisite for its success is the precise intraoperative localization of all canal orifices on the pulp chamber floor under the dental operating microscope, as these orifices serve as entry points to the root canal system [1].

In clinical practice, however, identifying all canal orifices is a frequent challenge, largely due to the presence of *concealed* orifices [2]. Among these, the second mesiobuccal canal (MB2) in maxillary molars is the most notoriously missed. Located within the mesiobuccal root, the MB2 orifice is typically situated palatal or slightly mesial to the main mesiobuccal (MB) canal. Its detection is severely hampered by factors like anatomical variations, dentinal overgrowth, and especially, calcification. Calcific deposits not only physically obstruct the orifice but also mimic the color and texture of the surrounding dentin, making it nearly indistinguishable even under a dental operating microscope [3]. Consequently, missed MB2 canals are strongly associated with persistent periapical lesions and may nearly quadruple the risk of treatment failure [4], making their detection critically important.

Given the difficulty of visually identifying MB2 under calcified or morphologically complex conditions, clinicians often turn to advanced technologies for support [5]. While the dental operating microscope provides essential magnification, pre-operative aids like Cone-Beam Computed Tomography (CBCT) are often used to create a 3D "map" to guide the intraoperative search [6]. However, even with these combined tools, the outcome remains heavily reliant on the clinician's subjective interpretation and clinical experience. Mentally registering 3D CBCT volumes with real-time 2D microscope views imposes a significant cognitive burden and demands advanced spatial reasoning, skills that may not be well developed in general practitioners or early-career endodontists. This reliance introduces considerable inter-operator variability and diagnostic inconsistency.

Given these persistent challenges, there is a pressing need for a computer-aided detection (CAD) framework. Such a framework could serve as an intelligent, objective, and standardized tool to assist clinicians in identifying concealed canal orifices, particularly MB2, under the microscope view. It would reduce operator variability, alleviate cognitive load, and enhance diagnostic accuracy, thereby reducing the risk of missed canals in routine endodontic procedures.

To achieve this goal, we propose an easy-to-deploy and efficient CAD framework for identifying concealed MB2 canal orifices. In contrast to prior approaches that rely solely on CBCT for canal detection [7], our method leverages microscope-view RGB images as the primary input modality. To the best of our knowledge, this is the first work to perform MB2 canal orifice detection directly from dental microscope images, which better aligns with real-time intraoperative scenarios. Specifically, we design a novel two-step pipeline. First, a dedicated segmentation network is introduced to detect canal orifices from RGB images, featuring a multi-scale, multi-head attention encoder and a hierarchical difference propagation decoder. Then, to further improve localization accuracy and mimic the real-world diagnostic workflow, we incorporate CBCT-derived anatomical priors in a post-processing step to calibrate and refine the initial predictions. This integration of both RGB and CBCT modalities not only enhances diagnostic precision but also simulates the multi-source information fusion process typically employed by clinicians, thereby improving the framework's clinical interpretability and applicability. Our main contributions are as follows:

- We propose an effective CAD framework for detecting concealed root canal orifices, with a focus on MB2. To the best of our knowledge, this is the

first work to perform orifice detection directly from RGB microscope images, rather than relying solely on CBCT.
- We introduce a CBCT-guided post-processing strategy to calibrate the detection results, which not only enhances spatial accuracy but also simulates real clinical workflow, thereby improving the interpretability and trustworthiness of the framework.
- We conduct extensive experiments on a private, clinically collected dataset. The results demonstrate the effectiveness of our method, especially in identifying MB2 orifices under challenging visual conditions.

2 Related Work

Recent studies [8–11] have increasingly applied deep learning to dental analysis. However, its application to microscope-view orifice detection remains unexplored. Therefore, our review focuses on the most relevant existing applications, which have predominantly utilized Cone Beam Computed Tomography (CBCT) imaging to analyze dental anatomy.

Deep Learning in Canal Orifice Detection. Existing literature focuses predominantly on CBCT-based pulp cavity and root canal segmentation. For example, Gamal et al. [12] introduced the Pulpy3D dataset and proposed a semantic segmentation framework that jointly delineates the pulp cavity, root canal systems, and the inferior alveolar nerve from 3D CBCT scans. Fontenele et al. [7] developed a CNN-based tool specifically for automated root canal segmentation in single-rooted teeth, achieving high accuracy on CBCT data. Additionally, Wang et al. [13] proposed a deep multi-task learning framework for joint tooth and root canal segmentation from CBCT, facilitating endodontic treatment planning by leveraging shared representations.

3 Method

We formulate orifice localization as a semantic segmentation task to achieve pixel-level delineation, which is better suited for handling the small size, irregular shape, and ambiguous boundaries of the targets. Our overall pipeline (Fig. 1) employs a two-stage approach: an orifice detection network first generates an initial segmentation from the 2D microscope view, which is then refined by a post-processing module leveraging pre-operative CBCT data. The detailed network architecture is presented in Fig. 2.

3.1 Orifice Detection Network Architecture

The core of our framework is a deep learning model featuring an encoder-decoder architecture, specifically designed to handle the challenges of microscope-view imagery. The encoder extracts robust multi-scale features, which are then progressively refined by the decoder to produce a pixel-level segmentation map.

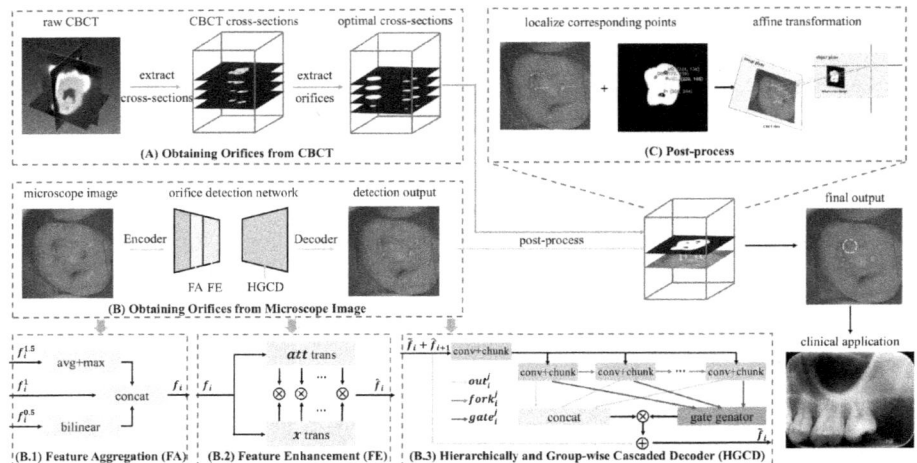

Fig. 1. Overview of our hybrid framework for concealed orifice detection. The pipeline generates an initial prediction from an intraoperative 2D microscope image (B), which is then refined in a post-processing step (C). This refinement is achieved by registering the prediction with pre-operative orifice locations extracted from 3D CBCT scans (A) via an affine transformation.

Encoder Module. The encoder is designed to capture scale-specific information from the input. It begins by creating a multi-scale representation of the input microscope image by resizing it to 1.5x, 1x (original), and 0.5x, a common zoom strategy in existing works [14]. These three scaled inputs are fed through a shared backbone network to generate feature maps ($f_i^{1.5}$, f_i^1, $f_i^{0.5}$) at multiple stages. This allows the network to capture both fine-grained local details from the upscaled view and global contextual information from the downscaled view. At each stage i: these multi-scale features are processed by two key blocks. First, a **Feature Aggregation (FA)** block aligns and merges the features. The 1.5x scale feature map ($f_i^{1.5}$) is downsampled using hybrid average-max pooling, while the 0.5x scale map ($f_i^{0.5}$) is upsampled via bilinear interpolation. These are then concatenated with the original 1x scale feature map to produce a single tensor f_i. Next, the aggregated f_i are passed to a **Feature Enhancement (FE)** block, which employs a multi-head attention mechanism. The final output $\hat{f}_i = \text{Concat}(\hat{f}_{i,1}, \hat{f}_{i,2}, \ldots, \hat{f}_{i,G})$ is formed by first computing an enhanced feature $\hat{f}_{i,g}$ for each of the G parallel groups, and then concatenating their outputs:

$$\hat{f}_{i,g} = \sum_{j=1}^{3} att_g^{(j)}(f_i) \odot x_g^{(j)}(f_i), \tag{1}$$

where $\odot$ denotes element-wise multiplication, and $att_g^{(j)}(\cdot)$ and $x_g^{(j)}(\cdot)$ are the attention weights and content features for scale j in group g, respectively.

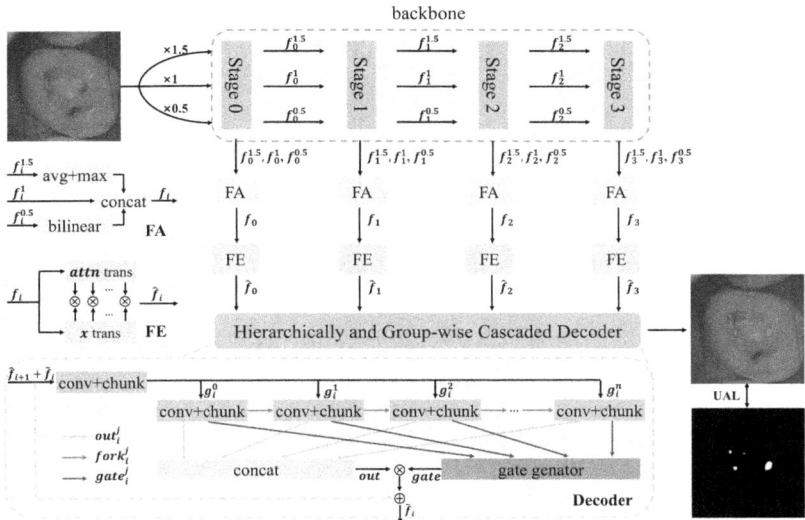

Fig. 2. The detailed architecture of our orifice detection network

Decoder Module. The decoder, termed the **Hierarchically and Group-wise Cascaded Decoder (HGCD)**, progressively refines the feature maps to generate the final segmentation. At each stage i, it first fuses the feature map from the encoder $\hat{f}_i$ with the upsampled output from the deeper stage $\tilde{f}_{i+1}$ to create the block's input $f'_i = \hat{f}_i + \text{Up}(\tilde{f}_{i+1})$, where $\text{Up}(\cdot)$ denotes a bilinear upsampling operation. Within the HGCD block, f'_i is split into multiple groups that interact in a cascaded manner: $fork$ connections pass information between groups for iterative refinement, while a $gate$ generator creates a channel-wise attention vector $\psi(\cdot)$ from all $gate$ branches:

$$\psi(\text{gate}_i) = \sigma(\text{Linear}_2(\text{ReLU}(\text{Linear}_1(\text{GAP}(\text{gate}_i))))), \qquad (2)$$

where σ denotes the Softmax activation function, $\text{GAP}(\cdot)$ is the global average pooling, and $\text{Linear}_{1,2}$ are two linear layers. This vector then modulates the concatenated out branches. The final output $\tilde{f}_i$ is obtained via a residual connection:

$$\tilde{f}_i = \text{Conv}(out_i \odot \psi(gate_i)) + f'_i. \qquad (3)$$

This iterative refinement process allows the network to meticulously delineate the complex boundaries of the canal orifices.

Loss Function. Following prior works [14], we use a combination of the standard binary cross-entropy (BCE) loss and an Uncertainty-Aware Loss (UAL). Secifically, given an input image I, the network outputs a prediction map $P \in [0,1]^{H \times W}$, where H and W are the image dimensions. Let $G \in [0,1]^{H \times W}$

be the corresponding binary ground-truth mask. The BCE loss is formulated as:

$$\ell_{BCE} = -\frac{1}{H \times W} \sum_{h=1}^{H} \sum_{w=1}^{W} [G(h,w) \log(P(h,w)) + (1 - G(h,w)) \log(1 - P(h,w))]. \tag{4}$$

Due to the ambiguous boundaries of canal orifices, training with BCE loss alone can lead to predictions with high uncertainty (i.e., pixel values close to 0.5). To address this, we incorporate a UAL to penalize such fuzzy predictions and compel the model to produce a more decisive output as follows:

$$\ell_{UAL} = -\frac{1}{H \times W} \sum_{h=1}^{H} \sum_{w=1}^{W} \left(1 - |2P(h,w) - 1|^2\right). \tag{5}$$

Then the final objective function is a weighted sum of the BCE and UAL losses:

$$\ell_{total} = \ell_{BCE} + \lambda \cdot \ell_{UAL}. \tag{6}$$

Here, λ is a balancing coefficient.

3.2 CBCT-Guided Post-Processing

In this section, we introduce a final post-processing stage that refines the network's output using pre-operative CBCT data to simulate the real clinical workflow. This stage simulates the clinical workflow of using a 3D map to guide a 2D search, effectively mitigating the risk of missed orifices like MB2. The core of this process is a 2D affine transformation, computed from three corresponding landmark orifices (Palatal, Mesiobuccal, and Distobuccal). To ensure a robust registration, these correspondences are not assumed but actively identified:

1. Landmarks (P, MB, DB) are first automatically located in both the network's prediction and a set of candidate CBCT slices using pre-defined anatomical rules (e.g., area size, relative distances).
2. The optimal CBCT slice is then determined by finding the best structural match between the P-MB-DB triangle in the prediction and those in the CBCT candidates.

Once the optimal correspondence and transformation are established, the matrix is used to project the location of the concealed MB2 orifice from the selected CBCT slice onto the final prediction map. This step corrects for potential omissions by the network. Additionally, an iterative merging strategy is employed to correct for over-segmentation by removing small, spurious detections, ensuring the final output is anatomically plausible.

4 Experiments

4.1 Experimental Settings

Dataset. Our study is based on a custom Canal Orifice Dataset collected intraoperatively from maxillary molars. The data was collected between January 2023 and March 2025 from the Department of Oral and Maxillofacial Surgery, Affiliated Stomatology Hospital of Tongji University. The dataset used in our experiments consists of 556 images, all from cases confirmed to contain the second mesiobuccal (MB2) canal. All images were captured after pulp chamber opening but before instrumentation.

Table 1. Performance Comparison of Different Methods on the Canal Orifice Dataset

Methods	standard metrics		clinically informed metrics				
	mIoU↑	DSC↑	ODR↑	PCR↑	MB2-DR↑	MB2-MR↓	mLE (px)↓
UNet	0.7115	0.5830	0.6475	0.2017	0.4118	0.5882	11.47
nnUNet	0.7492	0.6587	0.7723	0.3782	0.5462	0.4538	11.72
BiRefNet	0.6566	0.4613	0.3029	0.0168	0.0588	0.9412	10.54
SINetv2	0.7546	0.6698	0.8908	0.5714	0.7143	0.2857	12.14
Ours	0.7710	0.6937	0.9353	0.7563	0.8151	0.1849	**10.23**
Ours-pp	**0.7750**	**0.7006**	**0.9521**	**0.8992**	**0.8908**	**0.1092**	10.44

Implementation Details. We conducted all experiments using PyTorch on an NVIDIA GeForce RTX 4090 GPU. The network's encoder was initialized with weights from a PVT [15] model pre-trained on ImageNet, while all other layers were initialized randomly. We used the Adam optimizer with an initial learning rate of 1e-4, which decayed following a step strategy. The model was trained for 150 epochs with a batch size of 4, and all input images were resized to 384 × 384 during training.

Evaluation Metrics. To provide a comprehensive assessment, we evaluated performance using both standard segmentation metrics and custom, clinically-informed metrics developed in collaboration with endodontic experts. For pixel-level accuracy, we used the **Mean Intersection over Union (mIoU)** and **Dice Similarity Coefficient (DSC)**. To measure clinical utility, we introduced five instance-level metrics. **Orifice Detection Rate (ODR):** The percentage of correctly detected orifices out of the total ground-truth orifices. **Perfect Case Rate (PCR):** The percentage of images where all orifices were correctly identified with no false positives. **MB2 Detection Rate (MB2-DR) and Miss Rate (MB2-MR):** Metrics specifically evaluating the sensitivity and risk of missing the critical MB2 canal. **Mean Localization Error (mLE):** The average

Euclidean distance (in pixels) between the centers of matched predictions and ground-truth orifices. A detection was considered successful if the predicted center was within 50% of the ground-truth orifice's diameter.

Comparative Methods. To the best of our knowledge, this is the first work to address root canal orifice detection directly from intraoperative microscope RGB images. As there are no prior methods for direct comparison, we selected several state-of-the-art (SOTA) models from relevant domains to rigorously validate the effectiveness of our proposed framework. The selected baselines include **UNet** and **nnUNet**, which are benchmarks in medical image segmentation, as well as **BiRefNet** [16] and **SINetv2** [17], which are powerful models in Camouflaged Object Detection, a task that shares challenges with our own.

4.2 Performance Comparison

Quantitative Results. The quantitative comparison on our main clinical dataset is presented in Table 1. The results clearly show that our proposed network (Ours) significantly outperforms all baseline methods across both standard and clinically-informed metrics. For instance, our method achieves the highest mIoU of 0.7710 and a PCR of 75.63%. The introduction of our CBCT-guided post-processing (Ours-pp) provides a substantial further improvement, boosting the PCR to an impressive 89.92% and the critical MB2 Detection Rate to 89.08%, demonstrating its powerful capability to correct errors and complete the detection of the most frequently missed canal.

Qualitative Analysis and Robustness. Figure 3 provides a qualitative comparison on paired clinical cases (caseA) and their simulated calcified counterparts (caseB) to demonstrate our method's accuracy and robustness. The visualizations clearly show our method's superiority. While baseline methods like BiRefNet and nnUNet frequently miss orifices even in standard clinical views (e.g., caseA-1), their performance degrades drastically on the challenging calcified versions where they often fail completely (e.g., caseB-3). In contrast, our network (Ours) reliably identifies all concealed orifices in these difficult scenarios. The final results (Ours-pp) further showcase how our CBCT-guided post-processing refines the predictions for optimal accuracy, confirming our framework's effectiveness in clinically challenging situations.

4.3 Ablation Study

To validate the contribution of main components, including Feature Aggregation (FA), Feature Enhancement (FE), Hierarchically and Group-wise Cascaded Decoder (H), Uncertainty-Aware Loss (U) and CBCT-Guided Post-processing (C), we conducted a series of ablation experiments, with the results summarized in Table 2. The study demonstrates the effectiveness of our design, as removing any single component from our full framework leads to a clear degradation in performance across all metrics.

Automatic Concealed Orifice Detection from Microscope Imagery   61

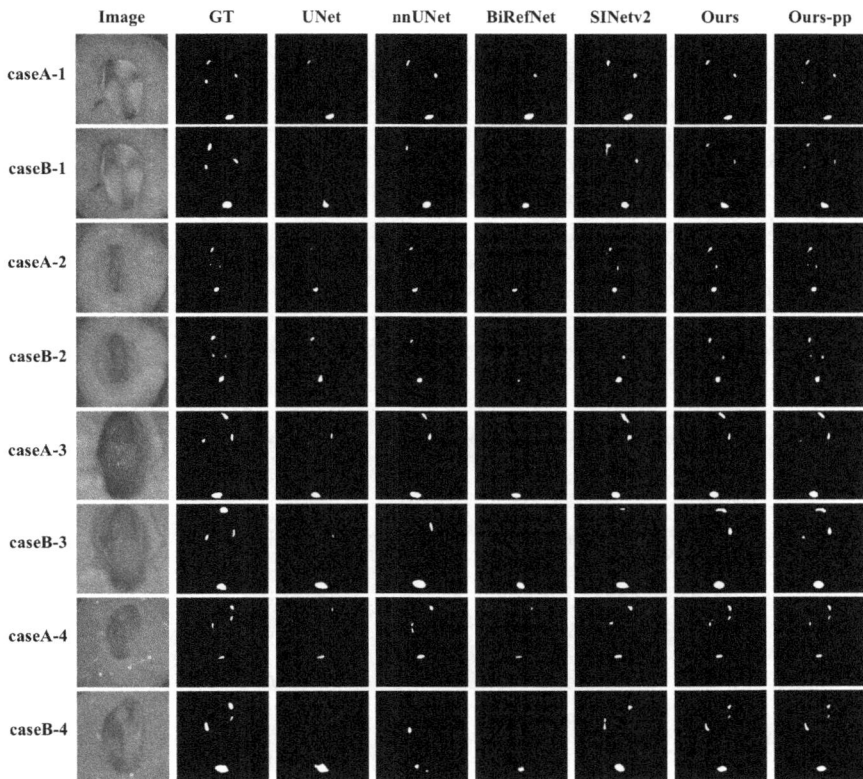

Fig. 3. Visualization of Segmentation Results on the Canal Orifice Dataset and its Simulated Calcified Counterparts. The figure shows a qualitative comparison on four clinical cases (A1-A4) and their corresponding simulated calcified versions (B1-B4). The calcified counterparts were created by injecting flowable composite resin around the orifice area to effectively mimic the visual characteristics of calcified tissue.

Table 2. Ablation Studies on Main Components

Ablation					standard metrics		clinically informed metrics				
FA	FE	H	U	C	mIoU↑	DSC↑	ODR↑	PCR↑	MB2-DR↑	MB2-MR↓	mLE (px)↓
	✓	✓	✓		0.7293	0.6210	0.6916	0.2101	0.4706	0.5294	12.71
✓		✓	✓		0.7236	0.6059	0.6210	0.1681	0.3445	0.6555	11.79
✓	✓		✓		0.6980	0.5539	0.5723	0.1092	0.2605	0.7395	12.89
✓	✓	✓			0.7102	0.5817	0.7046	0.3782	0.5378	0.4622	12.48
✓	✓	✓	✓		0.7710	0.6937	0.9353	0.7563	0.8151	0.1849	**10.23**
✓	✓	✓	✓	✓	**0.7750**	**0.7006**	**0.9521**	**0.8992**	**0.8908**	**0.1092**	10.44

4.4 Clinical Case Studies

To further demonstrate the practical effectiveness of our approach in clinically challenging scenarios, we present two representative cases in Fig. 4: a tooth with a severely calcified second mesiobuccal (MB2) orifice (Case I) and another with visually indistinct orifices (Case II). In both scenarios, the initial prediction from our deep learning network alone is insufficient, failing to localize the concealed MB2 canal (B). However, after applying our CBCT-guided post-processing, the missing orifice is accurately recovered in the refined result (C). The clinical validity of this refinement is confirmed by the successful treatment outcomes shown in the postoperative radiographs (D). These cases underscore the essential role of our hybrid 2D-3D approach in overcoming difficult clinical scenarios and highlight its potential as a reliable intraoperative assistant.

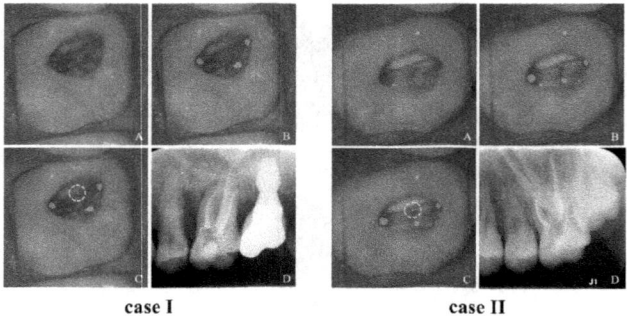

case I   case II

Fig. 4. Clinical Case Analysis on Challenging Maxillary Molars. Each case includes: (A) Intraoperative microscope view after access preparation, (B) predicted segmentation results, (C) post-processed results incorporating CBCT, and (D) postoperative radiograph. Case I illustrates calcified MB2; Case II features visually indistinct orifices.

5 Conclusion

In this paper, we proposed OrificeNet, a novel framework to address the critical challenge of detecting concealed root canal orifices from intraoperative microscope images. Extensive experiments demonstrate that our framework significantly outperforms state-of-the-art methods, showing strong robustness on challenging clinical and synthetic data. Ultimately, OrificeNet presents an effective and clinically-translatable solution to reduce the risk of missed canals and enhance the success rate of root canal treatments.

Acknowledgments. This work was supported by the National Natural Science Foundation of China (No. 62173252, 62476165)

Disclosure of Interests. The authors have no competing interests to declare that are relevant to the content of this article.

References

1. Erdemir, A., Ari, H., Güngüneş, H., Belli, S.: Effect of medications for root canal treatment on bonding to root canal dentin. J. Endod. **30**(2), 113–116 (2004)
2. Baruwa, A.O., et al.: The influence of missed canals on the prevalence of periapical lesions in endodontically treated teeth: a cross-sectional study. J. Endod. **46**(1), 34–39 (2020)
3. Chaniotis, A., Ordinola-Zapata, R.: Present status and future directions: management of curved and calcified root canals. Int. Endod. J. **55**, 656–684 (2022)
4. Bauman, R., Scarfe, W., Clark, S., Morelli, J., Scheetz, J., Farman, A.: Ex vivo detection of mesiobuccal canals in maxillary molars using cbct at four different isotropic voxel dimensions. Int. Endod. J. **44**(8), 752–758 (2011)
5. Essam, O., Boyle, E., Whitworth, J., Jarad, F.: The endodontic complexity assessment tool (e-cat): a digital form for assessing root canal treatment case difficulty. Int. Endod. J. **54**(7), 1189–1199 (2021)
6. Martins, J.N., Marques, D., Silva, E.J.N.L., Caramês, J., Versiani, M.A.: Prevalence studies on root canal anatomy using cone-beam computed tomographic imaging: a systematic review. J. Endod. **45**(4), 372–386 (2019)
7. Santos-Junior, A.O., Fontenele, R.C., Neves, F.S., Tanomaru-Filho, M., Jacobs, R.: A novel artificial intelligence-powered tool for automated root canal segmentation in single-rooted teeth on cone-beam computed tomography. Int. Endod. J. (2025)
8. Ren, G., Chen, Y., Qi, S., Fu, Y., Zhang, Q.: Feature patch based attention model for dental caries classification. In: Chen, Y., et al. (eds.) Clinical Image-Based Procedures. CLIP 2022. LNCS, vol. 13746, pp. 62–71. Springer, Cham (2023). https://doi.org/10.1007/978-3-031-23179-7_7
9. Kim, G., Chen, Y., Qi, S., Fu, Y., Zhang, Q.: Uncertainty based border-aware segmentation network for deep caries. In: Wesarg, S., et al. (eds.) Clinical Image-Based Procedures, Fairness of AI in Medical Imaging, and Ethical and Philosophical Issues in Medical Imaging. CLIP EPIMI FAIMI 2023 2023 2023. LNCS, vol. 14242, pp. 70–80. Springer, Cham (2023). https://doi.org/10.1007/978-3-031-45249-9_7
10. Qi, S., Fu, Y., Shan, H., Ren, G., Chen, Y., Zhang, Q.: Localisation and classification of multi-stage caries on cbct images with a 3d convolutional neural network. Clin. Oral Invest. **29**(5), 246 (2025)
11. Shan, H., Chen, Y., Liu, W., Qi, S., Zhang, Q.: Semantic-aware synthesis network for dental caries image generation from cbct to micro-ct. In: 2024 IEEE International Conference on Bioinformatics and Biomedicine (BIBM), pp. 2376–2381 (2024)
12. Gamal, M., Baraka, M., Torki, M.: Automatic mandibular semantic segmentation of teeth pulp cavity and root canals, and inferior alveolar nerve on Pulpy3D dataset. In: Linguraru, M.G., et al. (eds.) Medical Image Computing and Computer Assisted Intervention – MICCAI 2024. MICCAI 2024. LNCS, vol. 15008, pp. 14–23. Springer, Cham (2024). https://doi.org/10.1007/978-3-031-72111-3_2
13. Wang, Y., et al.: Root canal treatment planning by automatic tooth and root canal segmentation in dental cbct with deep multi-task feature learning. Med. Image Anal. **85**, 102750 (2023)
14. Pang, Y., Zhao, X., Xiang, T.Z., Zhang, L., Lu, H.: Zoom in and out: a mixed-scale triplet network for camouflaged object detection. In: Proceedings of the IEEE/CVF Conference on Computer Vision and Pattern Recognition, pp. 2160–2170 (2022)
15. Wang, W., et al.: Pyramid vision transformer: a versatile backbone for dense prediction without convolutions. In: Proceedings of the IEEE/CVF International Conference on Computer Vision, pp. 568–578 (2021)

16. Zheng, P., et al.: Bilateral reference for high-resolution dichotomous image segmentation. arXiv preprint arXiv:2401.03407 (2024)
17. Fan, D.P., Ji, G.P., Cheng, M.M., Shao, L.: Concealed object detection. IEEE Trans. Pattern Anal. Mach. Intell. **44**(10), 6024–6042 (2021)

Ablate Them All: A Trajectory Planning for Concurrent Percutaneous Ablation of Multiple Tumors

Adela Lukes[1(✉)], Stefano Fogarollo[3], Reto Bale[1], and Wolfgang Freysinger[2]

[1] Interventional Oncology-Microinvasive Therapy, Department of Radiology, Medical University Innsbruck, Innsbruck, Austria
adela.lukes@i.med.ac.at, reto.bale@i-med.ac.at
[2] 4D Visualization Laboratory, ENT Clinic, Medical University Innsbruck, Innsbruck, Austria
[3] Interactive Graphics and Simulation Group, Department of Computer Science, University of Innsbruck, Innsbruck, Austria

Abstract. Radiofrequency ablation is a minimally invasive technique widely used for treating liver tumors, yet planning optimal probe trajectories for multiple tumors remains a significant challenge due to the complexity of avoiding critical structures and ensuring adequate tumor coverage. However, it is crucial to treat as many tumors as possible in one intervention to decrease patients' hospitalization time. Furthermore, utilizing a single probe for ablating multiple tumors on the same trajectory reduces the number of probes used, and consequently, the risk of complications, such as crossing critical structures or trajectory collisions. In this scenario, probes are advanced to the most distant tumor, and subsequent to conducting ablation, the probes are retracted to the proximal tumor for an additional ablation.

We propose a novel trajectory planning algorithm for ablation procedures, introducing an innovative multi-tumor planning strategy and force field-based navigation. Our genetic optimization algorithm is guided by a field derived from abdominal structures to enable efficient and safe navigation through complex anatomy.

A retrospective analysis, performed on 18 patients from our in-house dataset, with 1 to 4 tumors each, shows its usability in clinical scenarios. Our algorithm produces safe, non-colliding, and clinically compliant solutions for all cases in 5.7 min on average and achieves a mean coverage of 93.5% of tumors with 5 mm safety margin. Comparison on single-tumor cases with existing automated methods demonstrates the competitiveness of our algorithm. Furthermore, the method's ability to handle complex multi-tumor scenarios is a significant step toward clinical implementation.

Keywords: radiofrequency ablation · multifocal tumors · stereotaxy · 3D navigation · planning of multiple rigid trajectories · unified non-dominated sorting genetic algorithm

© The Author(s) 2026
M. Erdt et al. (Eds.): CLIP 2025, LNCS 16126, pp. 65–74, 2026.
https://doi.org/10.1007/978-3-032-05479-1_8

1 Introduction

Radiofrequency ablation (RFA) is a widely employed method to treat tumors in the liver or other solid organs. The method involves the insertion of stiff probes into the tumor and the ablation of tissue around the uninsulated tip. In the current clinical practice, the planning is done manually on a CT scan with 1 mm slice thickness, focusing primarily on complete coverage, avoidance of collisions with critical structures, and other trajectories. As the number of trajectories increases, the associated complexity and time associated with planning correspondingly rise. Consequently, in most clinics, RFA is applied exclusively to small tumors. Stereotactic radiofrequency ablation (SRFA) utilizes a navigation system to enable the treatment of complex cases, such as large or multiple tumors [2]. The technique can be combined with a pullback method, which facilitates the ablation of multiple lesions along a single trajectory. The probe is initially advanced to the most distant target, where ablation is performed. Subsequent lesions located along the same path are ablated during the retraction of the probe, reducing the number of insertions. Our clinic deeply exploits SRFA while treating tumors over 6 cm in diameter with ten trajectories and performing over 200 procedures annually [1].

Automatic trajectory planning algorithms for ablation procedures navigate multiple trajectories in the abdomen environment and, while adhering to the clinical constraints (e.g. not penetrating critical structures), aim for maximal coverage of the tumor and safety margin (in practice set to 5 mm). Since a single RFA probe has a 10 mm ablation radius, it is appropriate solely for small lesions with a diameter < 1 cm, and for larger lesions, multiple probes have to be used. In the following, we present related work focusing on the planning of multiple rigid trajectories for larger tumors. Villard et al. in 2005 investigated trajectory planning for RFA and proposed a method capable of generating two trajectories within approximately 10 min [13]. Li et al. employed a genetic algorithm NSGA-II [4] to optimize up to five trajectories with a 20 mm ablation zone radius [6]. Microwave ablation (MWA) is a closely related technique to RFA, with similar planning requirements; the principal difference lies in the size of the ablation zone, which is generally larger in MWA. Li et al. presented an approach for MWA enabling the computation of up to three trajectories [8]. Zhou et al. [15] proposed a genetic algorithm to determine optimal ablation trajectories, which they evaluated only on three cases without reporting the computation time. Liang et al. proposed a constraint-based method utilizing the pullback technique [9]. A heuristic-based method was introduced to enable rapid trajectory planning in combination with the pullback approach [7]. An analogous technique to RFA is cryoablation, which uses cooling of the tissue for tumor destruction. Automatic planning for this method was developed by [5] using bioheat simulation at the target points to determine the optimal probe placements. Trajectory planning methods have also been explored in the context of minimally invasive brain surgery [3], or deep brain stimulation for epilepsy patients [11].

2 Method

We propose a method based on a genetic algorithm (GA) for the simultaneous trajectory planning of multiple tumors. GA operates on a population of candidate solutions, evolving them through crossover and mutation to generate improved populations. In the context of trajectory planning, it explores the abdominal space by adjusting needle positions relative to critical structures (CSs), the liver, and the tumor. As this navigation step constitutes the computational bottleneck, it will be the primary focus of this study.

We use unified non-dominated sorting genetic algorithm III (UNSGA-III) [12] with reference directions in the objective space and tournament selection in crossover. The overall structure of our proposed algorithm is based on our previous study that focused on optimization for single tumors [10]. Similarly to our previous work, the search space is defined as the skin mesh and the tumors with a safety margin for entry and target points of each trajectory, without the need to predefine sets of such points. To extend the algorithm to multiple tumor scenario and maintain reasonable computation time, we create a repulsive and attractive field from CSs and the tumors. These fields are then used in the mutation operator to guide the trajectories.

We define seven reference directions as the eigenvectors of the objective space, each corresponding to one of the seven objective functions. The objective functions minimize the trajectory length, trajectory length in the liver (but keep the trajectory longer than length of the active tip of the probe - 3 cm), the angle between the trajectory and the normal of the skin and liver capsule, the number of trajectories, and maximize the tumor coverage and the length in the tumor (complying with the pullback technique). Moreover, we optimize the placement of entry points, aiming for a minimum separation of 10 mm, facilitating the placement of insertion guides. The constraints in our problem are defined as the collisions of the trajectories with CSs and each other, the tumor coverage, and a binary value indicating whether every tumor is treated. The last constraint is implemented to avoid scenarios in which a small tumor remains untreated due to the satisfactory coverage already provided by a larger tumor. The source code of our algorithm is provided at https://git.i-med.ac.at/hno/field-navigated-trajectory-planning/.

2.1 Abdominal Field Formulation and the Mutation Operator

The abdominal field is used to efficiently move and align the solution with the CSs and the tumor. CSs exert a repulsive field, steering the trajectories away to prevent collisions and avoid undesirable outcomes. In contrast, tumors generate an attractive field that guides the trajectories inward. The respective fields are computed as followss

$$R = K * \mathbf{I}, \tag{1}$$

$$F(I) = \text{sigmoid}\left(\frac{R}{s}\right), \tag{2}$$

$$G(I) = \nabla(F), \tag{3}$$

where I is the input image (binary, 1 where CS, 0 otherwise), K is the convolution kernel; a $10 \times 10 \times 10$ matrix of ones ($\mathbf{1}_{10\times10\times10}$), $*$ represents the convolution operation, and s is the scaling factor. For our problem setting, we set the scaling factor s to 100 to have a reasonable field strength. The convolution introduces a gradient within the binary image, as the values peak within the CSs and linearly decrease in the direction of the boundaries. Applying the sigmoid function to the convolved result produces a smooth field F, allowing the computation of the gradient G. Such gradient is zero at the centerline of CSs. However, in the algorithm, the resulting field force is computed along the entire probe trajectory. Combined with the highly asymmetrical nature of abdominal anatomy, this ensures that the vanishing gradient does not pose practical limitations.

In the mutation process, we consecutively modify the candidate. First, the algorithm calculates the coverage and the tumor parts that remain uncovered. In case of insufficient coverage, the algorithm creates new trajectories. We empirically set the threshold to 0.9, as this level of coverage allows the algorithm to adequately distribute trajectories for full tumor coverage while avoiding excessive probe usage. To generate new trajectories, we use two different setups. The first one creates new trajectory with raycasting, where from a randomly picked non-covered tumor point, we create multiple random lines to skin mesh and keep the non-colliding one. The second approach picks an existing trajectory close to the non-covered tumor volume and creates multiple trajectories around it in uniform spacing in 5–15 mm distance.

The remaining steps of the mutation algorithm focus on refining the trajectories. Initially, an attractive field generated by the tumors pulls each trajectory toward them. Afterward, inter-trajectory interactions are introduced: trajectories attract each other to improve coverage, but repel if they come closer than a predefined threshold, preventing collisions. Next, the trajectories are modified by applying the repulsive field induced by CSs. As last, the trajectories are reshaped to conform to clinical constraints, such as limiting their length to 150 mm; the length of the RFA probe.

2.2 Algorithm Workflow

Due to the presence of many local minima and a vast search space in the problem of trajectory planning of simultaneous ablation of multiple tumors, our method incorporates guiding strategies to efficiently navigate in the search space and achieve high-quality solutions quickly. We employ different optimization strategies based on the combined volume of all tumors in a given case. For smaller total volumes, all tumors are optimized simultaneously. In contrast, for larger volumes, a sequential strategy is adopted to ensure optimal coverage.

Algorithm 1. Sequential Optimization for Multiple Tumors

Input: T (Tumor set), e_{short} (Max. number of epochs for initial optimizations), e_{long} (Max. number of epochs for final optimization)
Output: S (Multiple solutions of possible trajectories)
1: $S \leftarrow$ empty initial solution
2: **if** if $|T| \leq 1$ or totalVolume(T) $\leq v_{thresh}$ **then**
3:      $S \leftarrow$ Optimize(T, S, e_{long})
4: **else**   ▷ Use sequential optimization
5:      Sort T by centroid-to-skin distance (deepest first)
6:      $T_{seq} \leftarrow$ empty set
7:      **for** t in T **do**
8:          Add t to T_{seq}
9:          **if** t is not the last tumor **then**
10:             $S \leftarrow$ Optimize(T_{seq}, S, e_{short})
11:          **else**
12:             $S \leftarrow$ Optimize(T_{seq}, S, e_{long})
13:          **end if**
14:      **end for**
15: **end if**
16: **return** S

Particularly, if the case includes multiple tumors with a combined volume exceeding 2 cm^3, we apply a sequential strategy that prioritizes deeper tumors first. Specifically, tumors are sorted by the distance of their centroids to the skin surface. The process, illustrated in Pseudocode 1, begins by optimizing the trajectories for the deepest tumor. Then, it continues iteratively, adding tumors to the optimization in descending order of depth. This aligns with the pullback technique, where a single probe can treat multiple tumors along its path. We start the optimization process with a limited number of maximum epochs allowed. In the final optimization stage, the algorithm runs longer on the full tumor set to refine the solutions, balancing quality and efficiency. Moreover, we propose a strategy for gradually increasing the maximum number of trajectories allowed throughout the optimization. We start with a reduced number, focusing the algorithm on navigating around CSs. If coverage using the current maximal number of trajectories remains suboptimal, the allowed maximum is increased.

3 Data

We used our in-house dataset from our university clinic with over 800 SRFA-treated patients. We evaluated the method on 18 recent cases with a total of 33 tumors. After anonymization, liver, tumors, and vasculature were manually segmented by a clinician and reviewed as sufficient by another clinician. Other abdominal structures were segmented automatically using [14]. The mesh processing steps are analogous as in [10]. Rib meshes were inflated inferiorly to reflect clinical practice of probe insertion superior to the next lower rib, minimizing the risk of penetration of arteries.

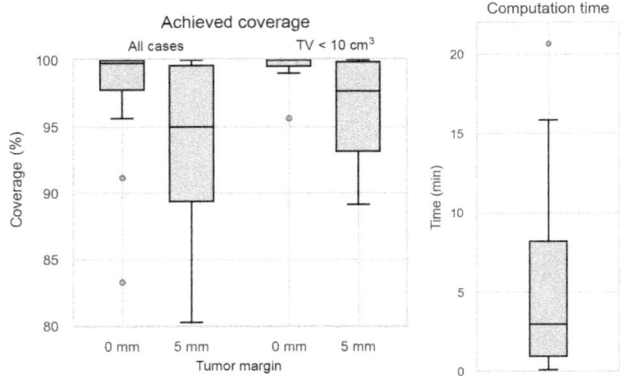

Fig. 1. Boxplots of achieved coverage (left) and computation time (right). Coverage results are presented with coverage and tumor margin on the x, y axis, for the entire dataset, as well as for a subset consisting of small tumors, defined as those with a volume less than < 10 mm^3. The right plot shows the computation time on the y-axis.

4 Evaluation

We evaluated the algorithm on Ubuntu with 8× Intel i7-9700K CPUs. Our algorithm proposed non-colliding solutions for all test cases and achieved tumor coverage of 93.5% ± 6.29 and 97.9% ± 4.17 with 5 and 0 mm margin. Figure 1 presents the achieved coverage and the computation time on our clinical dataset. The longest computation time was observed in the most complex case in our dataset; a patient with two tumors with a joined volume of 33 cm^3. The larger, 5 cm tumor was located subcapsularly at the junction of segments 2, 3, and 4; the second, 2.2 cm tumor was near a portal vein branch at the border of segments 4a and 5. Our algorithm achieved coverage of 80.3% (5 mm margin) and 83.3% (0 mm), compared to 77.9% and 81.9%, respectively, in the retrospective clinical plan. Due to insufficient initial coverage, additional trajectories were added during the procedure, highlighting the case's complexity.

Figure 2 demonstrates the comparison between the retrospective clinical and the automatic solution. The clinical solution achieved coverage of 73.9% and 92.1% of the tumor with 5 mm and 0 mm safety margin, respectively. Our algorithm produces a non-colliding solution in 6.2 min with coverage of 90.2% and 96.7%. The close alignment of the trajectories in the figure, along with the identical number of probes used in both the automatic and clinical plans (four), emphasizes the clinical relevance of our approach.

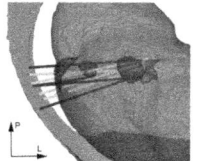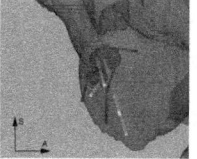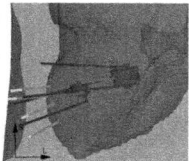

Fig. 2. Comparison between the automatic plan (blue lines) and the clinical plan (green lines) for two tumors treated using the pullback technique in top and side views. For clarity, only the tumors without safety margins (red, blue), the skin (magenta), and the liver (turquoise) are shown. Critical structures are omitted for clarity. Note that darker regions on the liver surface result from mesh rendering artifacts. (Color figure online)

4.1 Comparison with Previous Approach

To enable comparison with our prior method designed for single tumors [10], we select seven solitary tumors and apply both algorithms to these cases for evaluation of the performance.

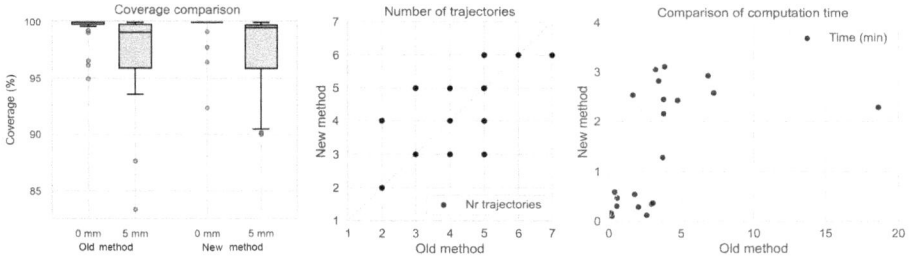

Fig. 3. Comparison of the previous and the new method in coverage, number of trajectories, and computation time. The dashed lines in the middle and right panels represent the identity line (y=x), indicating equal performance between the two methods. Note that in the last panel, the ranges of the x and y axes differ.

A visual comparison is presented in Fig. 3. The number of trajectories proposed by both methods was comparable. The mean achieved coverage was also similar, with values of 97.5% for the new method and 97.4% for the previous one. However, the new method achieved a higher median coverage and exhibited fewer outliers. The mean computational time for the previous approach was 3.25 ± 3.65 minutes, compared to 1.47 ± 1.15 min for the proposed method. In the right panel of Fig. 3, data points positioned to the right of the identity line indicate that the new method achieved faster computation times compared to the previous approach.

4.2 Comparison with Related Work

We compare our method on single tumor cases to recent work [8] for MWA planning using a 40 × 30 × 30 mm ablation zone ellipsoid. They evaluated performance on an in-house dataset, reporting tumor location and volume. For comparison, we selected single-tumor cases from our dataset with similar location and volume. Additionally, we indicate the proximity of CSs, such as vessels, that further increases the complexity of trajectory planning. The comparison of achieved coverage and computation time, together with trajectory features, is presented in Table 1.

Table 1. Comparison of our approach and automatic planning for MWA [8]. The first columns present the tumor location in Couinaud segments and immediate structures (IVC - inferior vena cava, PV(B) - portal vein (branch), D - diaphragm, P - pericardium, C - capsule), the tumor diameter (D), and volume (V). The first column in the Results section presents the coverage of tumors with 5 mm safety margin (C). For our method, we report the coverage for safety margins of 5 and 0 mm before and after slash. The last columns provide the number of probes (N), mean length of the trajectories (L), vertical deflection and liver normal angle within the corresponding CT slice for [8] while for us the mean trajectory angle to the normals of liver and skin, and lastly, the computation time (T).

	Test tumor				Results				
	Location	D (mm)	V (cm^3)		C (%)	N	L (mm)	Angle (°)	T (s)
[8]	1	23.2	4.68		99.4	2	119.0	2.5/0.3	19.6
Ours	1 (IVC,PV)	19	2.94		99.0/100	3	138.66	22.5	83.1
[8]	4	18.4	2.73		100	1	68.2	2.8/5.6	28.1
Ours	4 (D,P,C)	18	1.86		99.8/100	4	97.38	32.4	17.5
[8]	5	9.7	0.28		100	1	81.2	3.7/0.1	22.6
Ours	5	20	1.39		100/100	4	88.2	38.5	30.6
[8]	6	18.2	1.18		100	1	66.6	2.1/6.6	12.6
Ours	6 (C)	17.4	1.4		99.5/100	3	67.3	32.4	18.6

5 Discussion and Conclusion

We have proposed a planning algorithm for multi-tumor scenarios based on UNSGA-III, wherein the search is directed by fields generated by critical structures, tumors, and trajectories. The algorithm demonstrates high stability, with consistent convergence across multiple runs. Owing to its formulation, our approach for subsequent trajectory planning in cases of multiple large tumors has demonstrated efficacy in addressing complex scenarios and strategizing for the pullback procedure. Thus, it shows strong potential for clinical translation. The evaluation shows the consensus of our approach with clinical practice and its competitiveness with related work, despite optimizing additional trajectories due to optimizing with a smaller ablation zone. In addition, our method successfully plans trajectories for tumors in complex anatomical regions, including those near vascular structures or located deep within the liver; in the segment 1.

In future work, we aim to address the limitations of our algorithm in handling large tumors over 6 cm, which remain a challenging scenario. Moreover, we want to complete our algorithm with a software designed to assist physicians in their daily tasks, emphasizing a user-friendly and intuitive interface.

Acknowledgments. This research was funded in whole or in part by the Austrian Science Fund (FWF) 10.55776/DOC110. For open access purposes, the author has applied a CC BY public copyright license to any author accepted manuscript version arising from this submission.

Disclosure of Interests. The authors declare no conflict of interest.

References

1. Bale, R., Laimer, G., Schullian, P., Alzaga, A.: Stereotactic ablation: a game changer? J. Med. Imaging Radiat. Oncol. **67**(8), 886–894 (2023). https://doi.org/10.1111/1754-9485.13555
2. Bale, R., Widmann, G., Haidu, M.: Stereotactic radiofrequency ablation. Cardiovasc. Intervent. Radiol. **34**, 852–856 (2011). https://doi.org/10.1007/s00270-010-9966-z
3. De Momi, E., et al.: Automatic trajectory planner for StereoElectroEncephaloGraphy procedures: a retrospective study. IEEE Trans. Biomed. Eng. **60**(4), 986–993 (2012). https://doi.org/10.3390/s24165238
4. Deb, K., Pratap, A., Agarwal, S., Meyarivan, T.: A fast and elitist multiobjective genetic algorithm: NSGA-II. IEEE Trans. Evol. Comput. **6**(2), 182–197 (2002). https://doi.org/10.1109/4235.996017
5. Jaberzadeh, A., Essert, C.: Pre-operative planning of multiple probes in three dimensions for liver cryosurgery: comparison of different optimization methods. Math. Methods Appli. Sci. **39**(16), 4764–4772 (2016). https://doi.org/10.1002/mma.3548
6. Li, J., Gao, H., Shen, N., Wu, D., Feng, L., Hu, P.: High-security automatic path planning of radiofrequency ablation for liver tumors. Comput. Methods Programs Biomed. **242**, 107769 (2023). https://doi.org/10.1016/j.cmpb.2023.107769
7. Li, R., et al.: A heuristic method for rapid and automatic radiofrequency ablation planning of liver tumors. Int. J. Comput. Assist. Radiol. Surg. **18**(12), 2213–2221 (2023). https://doi.org/10.1007/s11548-023-02921-2
8. Li, S., et al.: Multi-stage automatic and rapid ablation and needle trajectory planning method for CT-guided percutaneous liver tumor ablation. Med. Phys. **52**(1), 113–130 (2025). https://doi.org/10.1002/mp.17450
9. Liang, L., Cool, D., Kakani, N., Wang, G., Ding, H., Fenster, A.: Multiple objective planning for thermal ablation of liver tumors. Int. J. Comput. Assist. Radiol. Surg. **15**(11), 1775–1786 (2020). https://doi.org/10.1007/s11548-020-02252-6

10. Lukes, A., Bale, R., Freysinger, W.: Automatic trajectory planning for stereotactic radiofrequency ablation in non-discrete search space. Inter. J. Comput. Assisted Radiol. Surgery 1–12 (2025). https://doi.org/10.1007/s11548-025-03386-1
11. Scorza, D., et al.: Retrospective evaluation and SEEG trajectory analysis for interactive multi-trajectory planner assistant. Int. J. Comput. Assist. Radiol. Surg. (11), 1–12 (2017). https://doi.org/10.1007/s11548-017-1641-2
12. Seada, H., Deb, K.: A unified evolutionary optimization procedure for single, multiple, and many objectives. IEEE Trans. Evol. Comput. **20**(3), 358–369 (2016). https://doi.org/10.1109/TEVC.2015.2459718
13. Villard, C., Soler, L., Gangi, A.: Radiofrequency ablation of hepatic tumors: simulation, planning, and contribution of virtual reality and haptics. Comput. Methods Biomech. Biomed. Engin. **8**(4), 215–227 (2005). https://doi.org/10.1080/10255840500289988
14. Wasserthal, J., et al.: TotalSegmentator: robust segmentation of 104 anatomic structures in CT images. Radiol. Artifi. Intell. **5**(5), e230024 (2023). https://doi.org/10.1148/ryai.230024
15. Zhou, Z., Qin, J., Ji, B., Jiang, Z., Zhao, J.: An automated ablation planning method for liver tumors. In: 2022 16th ICME International Conference on Complex Medical Engineering (CME), pp. 264–268 (2022). https://doi.org/10.1109/CME55444.2022.10063313

Open Access This chapter is licensed under the terms of the Creative Commons Attribution 4.0 International License (http://creativecommons.org/licenses/by/4.0/), which permits use, sharing, adaptation, distribution and reproduction in any medium or format, as long as you give appropriate credit to the original author(s) and the source, provide a link to the Creative Commons license and indicate if changes were made.

The images or other third party material in this chapter are included in the chapter's Creative Commons license, unless indicated otherwise in a credit line to the material. If material is not included in the chapter's Creative Commons license and your intended use is not permitted by statutory regulation or exceeds the permitted use, you will need to obtain permission directly from the copyright holder.

NEURAL: Attention-Guided Pruning for Unified Multimodal Resource-Constrained Clinical Evaluation

Devvrat Joshi and Islem Rekik(✉)

BASIRA Lab, Imperial-X (I-X) and Department of Computing,
Imperial College London, London, UK
i.rekik@imperial.ac.uk
http://basira-lab.com

Abstract. The rapid growth of multimodal medical imaging data presents significant storage and transmission challenges, particularly in resource-constrained clinical settings. We propose NEURAL, a novel framework that addresses this by using semantics-guided data compression. Our approach repurposes cross-attention scores between the image and its radiological report from a fine-tuned generative vision-language model to structurally prune chest X-rays, preserving only diagnostically critical regions. This process transforms the image into a highly compressed, graph representation. This unified graph-based representation fuses the pruned visual graph with a knowledge graph derived from the clinical report, creating a universal data structure that simplifies downstream modeling. Validated on the MIMIC-CXR and CheXpert Plus dataset for pneumonia detection, NEURAL achieves a 93.4–97.7% reduction in image data size while maintaining a high diagnostic performance of 0.88–0.95 AUC, outperforming other baseline models that use uncompressed data. By creating a persistent, task-agnostic data asset, NEURAL resolves the trade-off between data size and clinical utility, enabling efficient workflows and teleradiology without sacrificing performance. Our NEURAL code is available at https://github.com/basiralab/NEURAL.

Keywords: Multimodal Radiology Data · Vision-Language Models · Image Compression · Graph Neural Networks

1 Introduction

The volume of medical imaging data is expanding rapidly, with over two billion chest X-rays (CXRs) performed annually worldwide [1]. Each exam typically includes both an image and a radiology report, resulting in massive multimodal

---

GitHub: http://github.com/basiralab.

datasets that can reach terabytes in size. This data is vital for training AI models and supporting clinical workflows, but its scale creates serious challenges in real-world deployment. Many hospitals, especially in low-resource settings, face storage limitations, slow networks, and limited computing power [13]. These challenges delay timely image interpretation, making it difficult to integrate AI tools into clinical practice, and limit the reach of remote diagnostic services. Without effectively addressing these barriers, the full benefits of medical imaging AI may remain unrealized in real-world clinical settings [9].

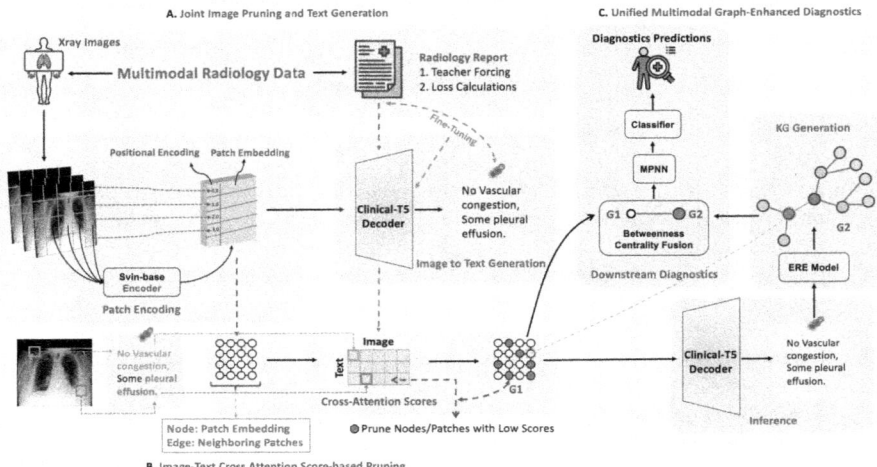

Fig. 1. End-to-end NEURAL pipeline for report-guided image pruning and graph-based clinical diagnostics.

Current efforts to manage this data complexity largely follow two distinct paradigms, each with fundamental limitations. The first is model-centric, focusing on accelerating computation. This ranges from established network pruning methods like CheXPrune [8] to more recent, sophisticated techniques that use language guidance to dynamically discard irrelevant tokens during inference [2,16]. While these approaches can significantly reduce the computational footprint of a model, their objective is transient model acceleration, not persistent data reduction. They still expect access to the original, full-resolution image for every task, thereby failing to alleviate the core issues of data storage and transmission. The second paradigm, conventional image compression, uses methods like JPEG to reduce file size but remains agnostic to clinical content, risking the degradation or loss of diagnostically critical details [5]. This presents a stark trade-off: one can either optimize the model while leaving the data logistics problem unsolved, or compress the data at the risk of compromising its clinical integrity.

To resolve this trade-off, we propose a novel framework centered on semantics-guided data compression. Our approach leverages the rich clinical narrative of the radiology report as a semantic blueprint to guide a targeted, structural compression of the associated image. Unlike methods that first downsize the image and risk losing information, our framework operates on the *full-resolution* image data, ensuring fine-grained visual details are considered during the pruning process. This identifies and preserves diagnostically critical regions while systematically discarding redundant information. The final output is not merely a compressed image with its radiological report, but a structured, multimodal representation in the form of a graph, which can be serialized into a lightweight format like a pickle file for efficient and lossless transmission.

This graph-based representation acts as a universal data structure that is inherently extensible. It is designed to seamlessly incorporate additional data types in the future, such as temporal clinical data or MRI scans, and to effectively model the complex interactions between them. Crucially, as we show in Sect. 3 (Parts B and C), this unification into a graph format eliminates the need for complex, task-specific models that handle heterogeneous inputs. Instead, a single, much simpler graph-based model can be applied for diverse downstream tasks, achieving comparable performance to other methods while operating on the highly compressed data.

Our pipeline operates on image text pairs from the MIMIC-CXR [7] and CheXpert Plus [3] dataset. Each image is first represented as a graph of visual patches (Refer Sect. 3-Part A). During fine-tuning, a ClinicalT5 [12] decoder is trained to generate the corresponding ground truth report from these visual inputs. The core innovation of our approach is to repurpose the cross-attention scores, calculated between full resolution image patches and clinical text tokens, as explicit signals for structurally pruning the image-graphs. We then evaluate the fidelity and clinical utility of the resulting compressed graph through two downstream tasks: (1) radiology report generation and (2) pneumonia classification. As a result, this work makes four key contributions:

1. We introduce the **first semantics-guided framework for radiological image compression** that uses cross modal attention from a generative vision language model to explicitly guide structural pruning of medical images, enabling highly targeted compression.
2. Our compression framework is **task-agnostic**, enabling the compressed data to be stored and utilized across downstream applications.
3. We develop a **rigorous dual-validation strategy**, using both report generation and disease classification to evaluate the fidelity and clinical relevance of the compressed representations.
4. Our framework is **designed for future extensibility**, leveraging a powerful betweenness centrality fusion to create a unified graph. This strategy is highly efficient, connecting modalities via a single semantically-meaningful link to avoid quadratic complexity , which in turn allows for the natural incorporation of new data types and simplifies downstream models.

2 Related Work

While NEURAL shares a core mechanism with contemporary language-guided pruning methods like LVPruning [16] and PuMer [2], their foundational objectives and task-dependencies diverge significantly. The primary goal of these other approaches is transient model acceleration, where pruning is intrinsically linked to a specific, concurrent downstream task. For instance, in an instruction-following model like IVTP [6] and ATP-LLaVA [18], the visual tokens are pruned based on the immediate query, making the pruning a consequence of that single task. NEURAL's objective, in contrast, is persistent data compression. It aims to create a smaller, permanent, and task-agnostic data asset that addresses the more fundamental challenges of data storage and transmission, particularly within resource-constrained clinical environments.

This distinction in purpose leads to a crucial divergence in methodology and generalizability. For other methods, the pruning is an ephemeral part of an inference pipeline, re-calculated for each new task or prompt. As a result, a different clinical application requires pruning of the full image specific to that task. NEURAL's methodology, however, is designed to be performed once; it uses the holistic clinical narrative of a ground-truth report to create a single, static, compressed graph. This resulting data asset is inherently versatile. Because the pruning is guided by the comprehensive report rather than a narrow downstream task, the compressed graph is a general-purpose representation that can be used for any number of subsequent clinical applications, be it pneumonia classification, report generation, or other diagnostic queries, without modification.

3 Methodology

Our framework proposes a semantics-guided, 3-stage approach to compress medical images and unify multimodal data into a single graph representation for downstream applications. The process begins by dividing a full-resolution chest X-ray into non-overlapping patches. Subsequently, a generative vision-language model is fine-tuned to create a radiology report, generating cross-attention scores that link text to the image patches. These scores are then repurposed to structurally prune the image, distilling it into a sparse graph containing only the most salient visual regions. Finally, this pruned visual graph is fused with a knowledge graph derived from the clinical report, creating the unified multimodal graph for efficient downstream diagnostics.

A) Joint Image Pruning and Report Generation. A central challenge in multimodal medical AI is bridging the semantic gap between the dense, low-level pixel data of a radiograph and the sparse, high-level concepts expressed in a clinical report. Our methodology addresses this by training an encoder-decoder module for a dual purpose: not only to generate coherent radiology reports but also, through this process, to produce a fine-grained alignment map identifying the most clinically salient image regions. This transforms the standard task of report generation into a tool for extracting the most important semantically aware visual regions in the image.

Our approach begins by processing a chest radiograph (CXR), $I \in \mathbb{R}^{H \times W \times 3}$. Following the Vision Transformer (ViT) paradigm [11], the image is divided into a sequence of N non-overlapping patches. These patches are linearly embedded, along with 2-dimensional positional embeddings for retaining global context, and fed into a Swin encoder [11]. This produces a sequence of patch-level feature representations, $V_{\text{img}} = \{v_1, v_2, \ldots, v_N\}$, where each $v_i \in \mathbb{R}^{D_{\text{vis}}}$ captures visual information from a specific image region. To make these features compatible with our language model, we project them into the text embedding space using a dedicated linear projection layer, resulting in the final visual embeddings $E_{\text{vis}} \in \mathbb{R}^{N \times D_{\text{text}}}$.

The core innovation lies in how we leverage these visual embeddings. We fine-tune a pre-trained clinical language model decoder (`Clinical-T5-Base`) [12] on report generation, conditioning it directly on the visual embeddings E_{vis}. During fine-tuning, we employ a teacher-forcing strategy: given an image I and its ground-truth report $R = \{t_1, t_2, \ldots, t_M\}$, the decoder learns to predict each token t_j based on the full set of visual embeddings E_{vis} and the preceding ground-truth tokens $\{t_1, \ldots, t_{j-1}\}$. This training encourages the model to establish direct, meaningful correlations between textual concepts and the specific image patches supporting them. A critical byproduct is the cross-attention mechanism, which produces attention scores at each decoding step quantifying the importance of each image patch for generating each token. Unlike conventional uses of attention for feature fusion, we repurpose this dynamic, context-aware attention map as a precise, data-driven signal to guide structural pruning of the visual graph, as detailed in the following section.

B) Image-Text Cross-Attention Score-based Pruning. In medical imaging, treating all parts of an image with equal importance creates a large volume of redundant data. This conventional method is inefficient and can obscure critical diagnostic clues within the noise. Our work introduces a new approach in radiological image analysis by leveraging the powerful semantic connection between an image and its medical report to intelligently filter this information. This process distills the dense image graph into a sparse, meaningful subgraph. By doing so, we create a focused map containing only the most clinically relevant visual evidence, leading to more efficient and accurate diagnostic models.

To facilitate this, the pruning mechanism repurposes the cross-attention scores generated during the report generation phase described previously. For each token t_j in the ground-truth report R, the decoder produces an attention weight vector, $\boldsymbol{\alpha}_j = \{\alpha_{j,1}, \alpha_{j,2}, \ldots, \alpha_{j,N}\}$, where the scalar $\alpha_{j,i}$ quantifies the importance of the i-th image patch, v_i, to generating that specific token. To determine the overall relevance of each patch to the entire clinical narrative, we aggregate these scores across all tokens. The cumulative importance score S_i for each patch v_i is computed as: $S_i = \sum_{j=1}^{M} \alpha_{j,i}$. This aggregated score S_i serves as a robust proxy for the clinical salience of the corresponding image region.

We define a threshold, τ, and construct a new, pruned set of vertices, V'_{img}, by retaining only those nodes whose cumulative attention scores exceed this threshold: $V'_{img} = \{v_i \in V_{img} \mid S_i > \tau\}$. The threshold τ can be determined empirically or set dynamically to retain a top-k percentage of the most salient patches, providing precise control over the desired level of compression. The resulting pruned graph, denoted as G_1 in Fig. 1, is a semantically compressed representation of the original image. It is no longer a generic grid of patches but a customized data structure shaped by the clinical text to retain only the most critical visual information, thereby reducing data size and focusing subsequent analyses on diagnostically significant regions.

C) Unified Multimodal Graph-Enhanced Diagnostics. The final stage of our framework aims for a holistic diagnostic capability, *moving beyond superficial feature fusion toward true structural integration of modalities*. We unify our pruned visual graph, G_1, with a structured representation of the clinical text. Simply using the raw report would overlook the rich interdependencies between medical concepts and image patches, and would require a multimodal model for handling text and graph simultaneously. We therefore transform the clinical narrative into a textual Knowledge Graph (KG), G_2. Using a BiomedVLP-CXR-BERT model used in [4], we extract medical findings and entities as nodes and represent their semantic relationships as edges. This structured representation allows the model to reason explicitly about how clinical concepts relate, rather than processing the report as a flat sequence of tokens.

We construct a knowledge graph G_2 from the report, and fuse it with the visual graph G_1, and then pass the combined structure through a MPNN for reasoning. The fusion is performed by connecting the nodes with the highest betweenness centrality from each graph, creating a semantically meaningful link between the two modalities. While other approaches, such as connecting all nodes based on cross-attention scores with edge weights, are possible, we choose to add only one edge to avoid the quadratic increase in complexity. This design supports efficient structural fusion, promoting interpretability and enabling a more context-aware multimodal diagnostic process.

To reason over this heterogeneous graph, we experimented with both homogeneous and heterogeneous graph neural network architectures. While heterogeneous GNNs preserve modality-specific semantics more explicitly, we found that their added complexity and training overhead did not yield performance gains significant enough to justify their use in this context. Instead, we adopt a standard Message Passing Neural Network (MPNN). Its iterative message-passing mechanism enables each node, whether an image patch or a clinical concept, to refine its representation based on both its neighborhood and its cross-modal links. This modeling choice strikes a practical balance between computational efficiency and representational richness.

At inference time, we consider two options: using the decoder to generate the text or directly using the textual report from the dataset. The choice depends on the specific requirements of the clinical setting, whether the hospital prefers to manage image-report pairs or rely solely on a compressed image graph for inference.

Why Graphs? Operating over a graph data structure offers significant advantages for both data storage and downstream tasks. First, it enables future extensibility by providing a unified framework that can naturally incorporate additional modalities, such as clinical temporal data and MRI, facilitating the modeling of interactions between heterogeneous data sources. Second, it standardizes inputs for downstream models, allowing them to work with a consistent graph format rather than a complex mixture of images, text, and temporal signals, thereby simplifying the input pipeline. Finally, the use of a cross-attention mechanism allows for efficient learning by projecting multimodal representations onto a shared low-dimensional manifold, enabling even simple message-passing neural networks (MPNNs) to effectively learn decision boundaries with relatively few parameters.

4 Experiments and Discussion

We conduct a comprehensive set of experiments to validate our proposed framework. Our evaluation is designed to answer three critical questions: (1) How effectively our method performs after data compression compared to established baselines? (2) Does the compressed graph representation retain sufficient clinical information for high-fidelity downstream tasks? (3) How does the quality of the textual guidance impact the trade-off between compression and diagnostic accuracy? We demonstrate that our approach achieves an unprecedented level of compression while maintaining state-of-the-art diagnostic performance, and we analyze the key factors influencing its behavior through extensive ablation studies.

Datasets. Our framework was rigorously evaluated using two distinct multimodal radiology datasets: MIMIC-CXR [7] and CheXpert Plus [3]. To ensure the integrity of our results and prevent data leakage, we enforced a strict patient-level separation, guaranteeing that only one imaging study per patient was included, thus preventing any patient's data from appearing in both training and testing splits.

The primary benchmark for classification was based on the MIMIC-CXR dataset, comprising 377,000 chest X-ray images. From this corpus, we identified 40,894 images with definitive pneumonia labels (24,338 negative and 16,556 positive), yielding a positive sample ratio of 40.5%. To better reflect the class imbalance typically seen in clinical practice, we constructed a more challenging dataset by sampling 10,000 of these labeled images to create a distribution with a pneumonia prevalence of only 15%. This allowed for a rigorous evaluation of model performance in detecting a sparsely represented target class. For external validation and assessment of generalization, we additionally utilized the CheXpert dataset. From its full set of 223,462 radiographs, we selected the 1,296 images that were explicitly labeled for pneumonia, providing a focused test set drawn from a distinct patient population. Finally, we divide the datasets into training (70%), validation (15%) and testing (15%) sets.

Baselines. We evaluated our framework for generating radiological reports using two other models. The RGRG [17] method detects anatomical regions in chest X-rays and generates region-specific sentences grounded on predicted bounding boxes, enhancing explainability and interactivity. CvT2DistilGPT2 [14] improves report generation by warm-starting its encoder with a Convolutional vision Transformer (CvT-21) pre-trained on ImageNet-21K and its decoder with DistilGPT2, resulting in more accurate and radiologist-like reports.

For the pneumonia detection task, we chose models that utilize both text as well as images for a fair comparison with NEURAL: CheXMed [15] is a multimodal algorithm for pneumonia detection that fuses features extracted from X-ray images via CNN and clinical notes processed through Named Entity Recognition into a combined representation for classification. RMT (Robust Multimodal Transformer) [10] assesses pediatric pneumonia severity by integrating X-rays and medical records using a Transformer architecture with multi-task learning and mask attention to handle missing data, achieving superior performance in multimodal settings.

Compression and Diagnostic Performance. Our framework strikes an effective balance between extreme data compression and high clinical accuracy, outperforming traditional approaches. As shown in Table 2, our method achieves a 97.7% reduction in data size by pruning the image graph down to just 2.3% on MIMIC-CXR dataset and 93.4% reduction in CheXPert dataset of its original nodes. Despite this significant compression, the AUC remains high at 0.947 and 0.875 for MIMIC-CXR and CheXPert datasets, surpassing the performance of other multimodal approaches that combine text and image inputs. While we initially use the radiology report generator to learn cross attention scores for pruning, it can also be leveraged to generate reports, eliminating the need to store original radiology texts. However, generating reports from pruned nodes does lead to a drop in performance, as reflected in lower AUC scores.

Table 1. BLEU-2 Scores for Radiology Report Generation.

Model	MIMIC-CXR	CheXpert	Model Params	Image Resolution
RGRG [17]	0.21	0.15	220M	512×512
CvT2DistilGPT2 [14]	0.09	0.08	102M	384×384
NEURAL (Ours)	**0.23**	**0.18**	308M	Full
NEURAL (Pruned)	0.19	0.16	308M	Pruned

Report Generation Quality. Although report generation is primarily used to extract cross attention scores for pruning, we also assess the quality of the generated reports to ensure the language model remains coherent. As shown in Table 1, our fine-tuned Clinical-T5 model achieves strong performance on the BLEU-2 metric, comparable to other models of similar size [14,17]. This suggests that our pruning strategy is guided by clinically meaningful and coherent text generation. In contrast to prior works that reduces image resolution, our patch-based method operates on full-resolution images, enabling the encoder to retain fine-grained visual details. We also evaluate Clinical-T5 on the reduced patch set, using a compression ratio similar to that in Table 2. While the BLEU-2 scores drop relative to the full data, the model still performs reasonably well, indicating some loss of coherence due to the pruned patches that have low cross-attention scores.

Table 2. Pneumonia Detection Performance (AUC) Across Datasets. CI references to Max-Compression on images, GT refers to the use of Generated Text. Results are defined as AUC vs % Compression

Model	MIMIC-CXR	CheXpert
CheXMed [15]	0.939, 0%	0.816, 0%
RMT [10]	0.915, 0%	0.869, 0%
NEURAL (No Pruning)	**0.963**, 0%	**0.902**, 0%
NEURAL (CI)	0.947, 97.7%	0.875, 93.4%
NEURAL (CI + GT)	0.891, 97.7%	0.838, 93.4%

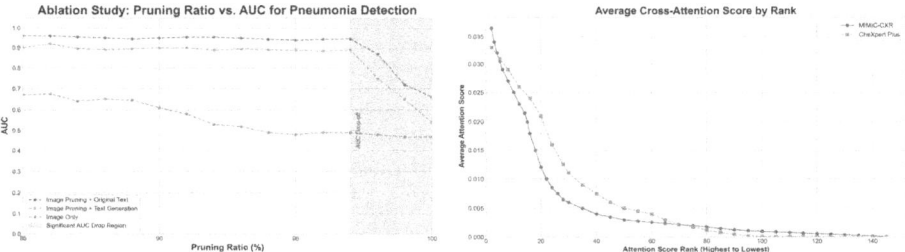

Fig. 2. Figure on the left shows the ablation study of varying the pruning ratio on three separate tasks for MIMIC-CXR dataset. Figure on the right shows the average cross-attention scores vs rank of the patch inside the image for the CheXpert and MIMIC-CXR datasets

Ablation Study. We analyze how varying the pruning ratio affects diagnostic accuracy. As shown in Fig. 2 (Left), retaining 2.3% of image patches already yields a strong AUC of 0.95. Increasing the retention to 10% results in only a slight gain (+0.01 AUC), indicating diminishing returns. This justifies our design choice: 97.7% compression delivers substantial efficiency with minimal impact on performance. Moreover, we demonstrated that removing text information entirely for prediction, combined with pruning over 90% of the data, leads to a substantial drop in performance. This highlights the critical role of text tokens in pneumonia detection. Since the other two models do not support unimodal input like NEURAL, we were unable to evaluate their performance using only image data.

To better understand why selecting the top 2.3% of patches is effective, we analyzed the cross-attention scores of the image patches. For each image, we ranked the patches in descending order based on their cross-attention scores. Then, for each rank, we computed the average score across all images and visualized the results in Fig. 2 (Right). The plot reveals that the top 20 ranked patches exhibit significantly higher attention scores, which then drop sharply to near zero. These 20 patches correspond to the top 2.3% patches across the MIMIC-CXR dataset. Similarly, in the CheXpert Plus dataset, a larger proportion of patches exhibit high attention scores, resulting in less aggressive compression compared to the MIMIC-CXR dataset.

5 Clinical Implications and Conclusion

Our method enables a 93–97% reduction in image size with minimal loss in task performance, offering significant advantages for clinical imaging workflows. This compression greatly reduces storage demands in Picture Archiving and Communication Systems (PACS) and supports efficient teleradiology in low-bandwidth or resource-constrained environments. The pruned image graphs can be stored in lightweight formats (e.g., pickle) and reused for a variety of downstream tasks. Since the pruning is driven by a task-agnostic report generation model, the approach remains generalizable and adaptable to multiple clinical applications. Unlike prior work that reduces image resolution, our patch-based method preserves full image detail, enabling better retention of fine-grained visual features.

In addition, our fine-tuned Clinical-T5 model can generate coherent and clinically meaningful radiology reports directly from the compressed inputs. This reduces the reliance on storing large, paired imagereport datasets and simplifies data management in clinical research and deployment settings. Although some drop in report quality is observed when using heavily pruned inputs, the model still performs reasonably well, demonstrating the effectiveness of our pruning strategy. Overall, this enables scalable report generation with minimal overhead. In future work, we aim to extend this framework to other data types, such as temporal or volumetric imaging, and to design models that can operate directly on compressed graph representations. This would move NEURAL toward a more efficient, generalizable, and end-to-end solution for multimodal clinical data analysis.

References

1. Akhter, Y., Singh, R., Vatsa, M.: AI-based radiodiagnosis using chest x-rays: a review. Front. Big Data **6**, 1120989 (2023). https://doi.org/10.3389/fdata.2023.1120989
2. Cao, Q., Paranjape, B., Hajishirzi, H.: PuMer: pruning and merging tokens for efficient vision language models. In: Rogers, A., Boyd-Graber, J., Okazaki, N. (eds.) Proceedings of the 61st Annual Meeting of the Association for Computational Linguistics (Volume 1: Long Papers), pp. 12890–12903. Association for Computational Linguistics, Toronto, Canada (2023). https://doi.org/10.18653/v1/2023.acl-long.721

3. Chambon, P., Delbrouck, J.B.: Chexpert plus: augmenting a large chest X-ray dataset with text radiology reports, patient demographics and additional image formats. arXiv preprint arXiv:2405.19538 (2024)
4. Delbrouck, J.B., Chambon, P.: RadGraph-XL: A large-scale expert-annotated dataset for entity and relation extraction from radiology reports. In: Ku, L.W., Martins, A., Srikumar, V. (eds.) Findings of the Association for Computational Linguistics: ACL 2024, pp. 12902–12915. Association for Computational Linguistics, Bangkok, Thailand (2024). https://doi.org/10.18653/v1/2024.findings-acl.765
5. Fischer, M., Neher, P., Schüffler, P., Xiao, S., Almeida, S.D., et al.: Enhanced diagnostic fidelity in pathology whole slide image compression via deep learning. In: MLMI 2023. MICCAI 2023, pp. 427–436. Springer, Cham (2023). https://doi.org/10.1007/978-3-031-45676-3_43
6. Huang, K., Zou, H., Xi, Y., Wang, B., Xie, Z., Yu, L.: IVTP: instruction-guided visual token pruning for large vision-language models. In: Leonardis, A., Ricci, E., Roth, S., Russakovsky, O., Sattler, T., Varol, G. (eds.) ECCV 2024, pp. 214–230. Springer, Cham (2025)
7. Johnson, A.E.W., et al.: MIMIC-CXR, a de-identified publicly available database of chest radiographs with free-text reports (2019). https://www.nature.com/articles/s41597-019-0322-0#citeas
8. Kaur, N., Mittal, A.: CheXPrune: sparse chest X-ray report generation model using multi-attention and one-shot global pruning. J. Ambient. Intell. Humaniz. Comput. **14**(6), 7485–7497 (2023). https://doi.org/10.1007/s12652-022-04454-z
9. Kulkarni, P., Kanhere, A.U., Siegel, E., Yi, P.H., Parekh, V.S.: One copy is all you need: resource-efficient streaming of medical imaging data at scale. arXiv preprint arXiv:2307.00438 (2023)
10. Li, J., et al.: Assessing severity of pediatric pneumonia using multimodal transformers with multi-task learning (2024). https://pmc.ncbi.nlm.nih.gov/articles/PMC11660274/
11. Liu, Z., Lin, Y., Cao, Y.: Swin transformer: hierarchical vision transformer using shifted windows. arXiv preprint arXiv:2103.14030 (2021)
12. Lu, Q., Dou, D., Nguyen, T.: ClinicalT5: a generative language model for clinical text. In: Findings of the Association for Computational Linguistics: EMNLP 2022, pp. 5436–5443. Association for Computational Linguistics, Abu Dhabi, United Arab Emirates (2022). https://doi.org/10.18653/v1/2022.findings-emnlp.398
13. Magudia, K., Bridge, C.P., Andriole, K.P., Rosenthal, M.H.: The trials and tribulations of assembling large medical imaging datasets for machine learning applications. J. Digit. Imag. **34**(6), 1424–1429 (2021). https://doi.org/10.1007/s10278-021-00505-7
14. Nicolson, A., Dowling, J., Koopman, B.: Improving chest x-ray report generation by leveraging warm starting. Artif. Intell. Med. **144**, 102633 (2023). https://doi.org/10.1016/j.artmed.2023.102633
15. Ren, H., et al.: CheXMed: a multimodal learning algorithm for pneumonia detection in the elderly. Inf. Sci. **654**, 119854 (2024). https://doi.org/10.1016/j.ins.2023.119854
16. Sun, Y., Xin, Y., Li, H., Sun, J., Lin, C., Batista-Navarro, R.: LVPruning: an effective yet simple language-guided vision token pruning approach for multi-modal large language models. arXiv preprint arXiv:2501.13652 (2025)

17. Tanida, T., Müller, P., Kaissis, G., Rueckert, D.: Interactive and explainable region-guided radiology report generation. In: 2023 IEEE/CVF Conference on Computer Vision and Pattern Recognition (CVPR), pp. 7433–7442. IEEE (2023). https://doi.org/10.1109/cvpr52729.2023.00718
18. Ye, X., Gan, Y., Ge, Y., Zhang, X.P., Tang, Y.: ATP-LLaVA: adaptive token pruning for large vision language models. arXiv preprint arXiv:2412.00447 (2024)

Interpreting CT-Scans with CLIP: An Explorative Study of Attribution Methods for 3D Vision-Language Models

David Avedis Injarabian[1], Joonas Ariva[1,2](✉), Hendrik Šuvalov[1], and Dmytro Fishman[1,2,3]

[1] University of Tartu, Tartu, Estonia
{david.avedis.injarabian,joonas.ariva,hendrik.suvalov,
dmytro.fishman}@ut.ee
[2] STACC, Tartu, Estonia
[3] Better Medicine, Tartu, Estonia

Abstract. Deep learning holds promise for supporting radiologists by addressing challenges such as high workloads, increasing imaging volumes, and inconsistencies in image interpretation. However, current models require extensive annotations to work efficiently. The annotation of 3D medical images demands substantial time and expert effort, restricting the scalability of clinical AI applications. This work explores whether radiology reports, which are readily available and semantically rich, can serve as weak supervision for medical image segmentation. We investigate a contrastive vision-language model trained to align 3D computed tomography (CT) scans with free-text reports and probe the resulting representations using interpretability techniques. By analyzing attribution patterns extracted from the model, we assess whether it captures spatially meaningful signals despite lacking segmentation labels. This approach aims to reduce reliance on manual annotations and move toward scalable, label-efficient segmentation pipelines. The resulting code and comprehensive 3D visualizations can be found at https://github.com/injardav/CT-CLIP-UT.

Keywords: CLIP · Medical Imaging · Weak Supervision · CT · VLP

1 Introduction

Deep learning has advanced medical image analysis, particularly in classification and segmentation tasks [1]. However, it still relies heavily on large annotated datasets, which is especially problematic for 3D imaging, where manual segmentation is labor-intensive and costly [2]. In contrast, radiology reports, which often accompany computed tomography (CT) scans, are routinely written by radiologists and offer a rich, underutilized source of weak supervision.

D. A. Injarabian, J. Ariva: These authors contributed equally to this work.

© The Author(s), under exclusive license to Springer Nature Switzerland AG 2026
M. Erdt et al. (Eds.): CLIP 2025, LNCS 16126, pp. 87–96, 2026.
https://doi.org/10.1007/978-3-032-05479-1_10

Recent de-identified datasets that pair 3D CT scans with free-text reports have made it feasible to explore multimodal learning at scale [3,4]. While vision-language models trained with Contrastive Language-Image Pretraining (CLIP) have shown strong generalization in natural image domains [5], their potential in clinical settings remains underexplored, especially in 3D modalities.

In this work, we study whether CLIP-style models can facilitate weakly supervised segmentation in 3D CT scans by leveraging textual supervision in the form of full reports. Although these models are trained to align entire reports with CT scans, we hypothesize that this alignment is primarily driven by pathology-related content, as it represents the most distinctive and case-specific information within the reports. To test this, we adapt a selection of widely-used attribution methods and apply them to a pretrained model to evaluate whether image-text alignment can reveal meaningful spatial structures in the scans and, more specifically, which attribution methods are suitable for this task.

In summary, our contributions are: (1) we investigate the use of CLIP-style vision–language models for weakly supervised segmentation in 3D CT scans by adapting and evaluating several attribution methods to assess image–text alignment; and (2) we release our code publicly to support reproducibility and further research.

2 Background

This section reviews the use of vision–language models in CT imaging and outlines attribution methods for interpreting CLIP-style models in general.

2.1 Vision-Language Pretraining

Vision-language pretraining (VLP) utilizes contrastive learning between image-text pairs to acquire generalizable features across both modalities in a weakly supervised manner [5]. The application of VLP in the medical domain is expanding rapidly [6]. However, the majority of existing research remains focused on X-ray imaging, primarily due to the greater availability of annotated datasets and the lower computational cost associated with 2D images compared to 3D volumetric images. Additionally, the inherent complexity of processing volumetric data further limits the adoption of CT scans in VLP research.

To mitigate these challenges, some VLP approaches for CT imaging use only small 3D patches of the scan, as demonstrated in CLIP-Lung [7]. However, this typically requires prior knowledge of relevant regions of interest (ROIs) in order to extract informative patches. This knowledge is often unavailable without additional annotations. Therefore, reliance on ROI annotations limits the applicability of such methods in fully weakly supervised settings. Similarly, CT-GLIP [8] reduces task complexity by relying on pre-existing organ segmentations and training a CLIP-based model on selected scan subsections.

This highlights the significance of CT-CLIP and CT-RATE, recently introduced by Hamamci et al. [3]. CT-CLIP is the first fully 3D transformer-based

VLP model for CT imaging, trained on CT-RATE, a unique dataset comprising approximately 25,000 chest CT scans paired with corresponding radiology reports. Due to these characteristics, CT-CLIP was selected as the foundational model for evaluating various attribution methods in our study.

2.2 Attribution Methods for CLIP-Type Models

CLIP demonstrates strong zero-shot capabilities, but interpreting its outputs, and deriving weakly supervised segmentations, relies on attribution methods, which have been developed primarily for natural images and for classification models. Therefore, some of these methods need modifications to work on CLIP-style models.

Attention-based attribution techniques, such as visualising self-attention maps or attention rollout [9], can easily be applied to CLIP image encoders. However, attribution methods, which rely on gradients or classification scores, usually need modifications to be adapted to CLIP style models, since CLIP outputs feature vectors and not class probabilities. Instead, a cosine similarity between the image and accompanying text prompt vector can be used in place of a class probability.

Some popular attribution methods have also had additional CLIP specific modifications. In recent CLIP-segmentation frameworks such as CLIP Is Also an Efficient Segmenter [10], Grad-CAM has been enhanced with techniques like softmax-over-attention, class-aware background sets, a real-time affinity module, and confidence-guided loss to produce high-quality pseudo-masks. Also Score-CAM [11] has had its CLIP specific implementation gScoreCAM [12], which uses gradients to rank feature maps and selects only the most influential for attribution map creation, speeding up the process and making it less noisy.

Despite the success of attribution methods in 2D vision tasks, none of these techniques, to the best of our knowledge, have been adapted for volumetric medical CLIP models. Our work addresses this gap by evaluating variety of common attribution methods (attention maps, attention rollout, integrated gradients, GradCAM, occlusion) on CT-CLIP, to assess their ability to extract weak segmentations maps of pathologies from 3D CT scans.

3 Methods

This section describes the vision-language model architecture used in our experiments and outlines the adaptations required to apply spatial attribution techniques to a contrastively trained model. Each method is described in the order it appears in the study, with emphasis on how it interfaces with the unique structure of CT-CLIP.

3.1 3D Vision-Language Model

Our experiments are based on CT-CLIP [3], a contrastively trained large vision-language model for volumetric chest CTs. CT-CLIP uses a volumetric Vision

Transformer CT-ViT [13] to encode 3D CT scans and a radiology-adapted BERT model [14] to encode full radiology reports. During training, the model learns to project paired image-text inputs into a shared embedding space using a symmetric contrastive loss. Importantly, CT-CLIP is not trained for segmentation or classification; it lacks token-level supervision and does not output class logits.

The CT-ViT encoder operates on non-overlapping 3D patches extracted from the input scan. Each patch is tokenized and passed through a hierarchical transformer. First, the patches are processed by a spatial transformer that encodes intra-slice relationships, followed by a causal transformer that captures inter-slice dependencies across the volume. The image encoder outputs a set of volumetric patch tokens, which are pooled and projected to yield a fixed-length image embedding. A similar process is applied to the input text using the language encoder. The cosine similarity between image and text embeddings serves as the models alignment score.

3.2 Attribution Methods

We apply five attribution methods to assess spatial alignment between visual inputs and text descriptions. Our hypothesis is that pathology-related regions in chest CT volumes significantly influence both the image and text feature representations produced by the model. Consequently, we expect that attribution methods can highlight these regions, offering a form of weak localization. Each method is adapted to produce 3D attribution maps over the input volume, using either internal model activations or gradient-based mechanisms. In all cases, similarity is computed between the image embedding and a fixed text embedding obtained from the full radiology report, and used in gradient-based methods.

Attention Maps. Attention weights from the final layer of the CT-ViT encoder are extracted to assess intra- and inter-slice token interactions. Attention scores from both the spatial and causal transformers are extracted and grouped by attention head and layer. Each attention matrix encodes token-to-token relevance within its respective branch. No aggregation or projection is performed beyond the encoder output; these maps are retained for head-wise and layer-wise comparison to investigate whether any interpretable or consistent spatial patterns emerge.

Attention Rollout. Attention rollout approximates the cumulative propagation of attention across transformer layers [9]. We compute rollout separately for the spatial and causal transformers of the CT-ViT model. Attention matrices are averaged across heads within each layer and combined recursively across layers using matrix multiplication, with residual connections approximated via identity weighting.

Unlike conventional attention rollout, which centers attention on a special classification token ([CLS]) used in classification models, CT-CLIP has no such token. Instead, attention scores are averaged across all spatial tokens in the final rollout matrix to produce a single 3D volume per transformer module. These are reshaped and upsampled to generate full-resolution attribution maps.

Integrated Gradients. Integrated gradients attribute voxel-level importance by computing gradients of the image–text similarity along a linear interpolation path between a baseline and the input scan [15]. The input baseline is a completely black CT volume (all voxels set to minimum intensity). A series of interpolated volumes is constructed along a linear path from the baseline to the original scan. At each step, the similarity score is computed, and its gradient with respect to the input image is recorded.

The final attribution map is obtained by averaging the gradients across steps and scaling by the input difference. This yields a voxel-wise relevance map reflecting how much each input voxel contributes to the image–text similarity score. The output is normalized and resized to match the input CT resolution.

Grad-CAM. Grad-CAM is adapted to the CT-CLIP architecture by targeting internal transformer activations [16]. We apply Grad-CAM to three different layers: both final layers of spatial and causal transformers and the final layer of the whole encoder—Vector Quantization (VQ) layer, which discretizes feature representations using a learned codebook. Gradients of the cosine similarity score are calculated with respect to the selected transformer activations, averaged spatially to obtain channel-wise weights, and used to compute a weighted sum of the feature maps. The result is passed through ReLU activation and normalized to produce an attribution map.

Occlusion Sensitivity. Occlusion sensitivity [17] is implemented by systematically masking small subvolumes of the scan and measuring the resulting change in image–text similarity. For each occluded region, we compute the cosine similarity between the modified image and the fixed text embedding. The decrease in similarity is treated as the attribution score for the masked region. Aggregating these scores across the scan yields a 3D attribution map indicating the relevance of different spatial regions.

3.3 Evaluation

Given the exploratory nature of this study and the absence of segmentation labels in the CT-RATE dataset, our evaluation is primarily empirical. We assessed the quality of the extracted 3D heatmaps through visual inspection, focusing on two main questions: (1) Do the methods concentrate attention in specific anatomical regions (lungs for chest CT dataset), or is the attention diffusely distributed across the entire scan? (2) Do these approaches show promise in the context of 3D medical imaging segmentation, or do they appear ill-suited for this task?

4 Results

We evaluated the spatial attribution behavior of the CT-CLIP model across five interpretability methods, assessing their ability to recover semantically meaningful and spatially localized activations from image–text alignment. Each method is applied to a set of held-out 3D CT scans from the CT-RATE dataset, using full radiology reports.

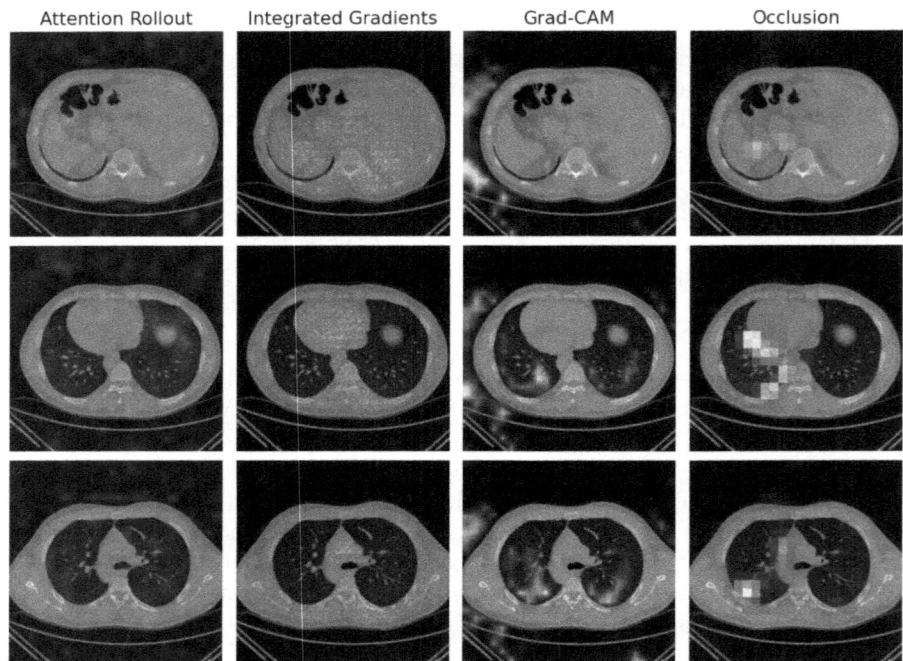

Fig. 1. Attribution methods applied to an example CT scan of a patient with lymphadenopathy and lung nodules. Representative axial slices are shown to illustrate model interpretations across the volume. We show the causal transformer variant of attention rollout and the VQ-layer Grad-CAM. Additional variants—such as spatial attention rollout and Grad-CAM applied to spatial and causal transformer layers—are available in the GitHub repository.

4.1 Qualitative Attribution Results

This section reports our qualitative findings in the order of method application. Visual examples of key results are provided in Fig. 1. Visualizations of full 3D attribution maps can be seen in the GitHub repository.

Attention Maps. We analyzed the attention weights of both spatial and causal transformers of CT-ViT. In both cases, attention values of a single head and layer appear noisy and do not correspond to identifiable anatomical structures. These findings suggest that attention weights from individual heads, when visualized directly, lack interpretable spatial localization.

Attention Rollout. The attention rollout maps from both spatial and causal transformers appear smooth and spatially contiguous. However, in both cases, the attribution is distributed broadly throughout the scan without anatomical specificity or alignment to known structures, resulting in diffuse and weak localizations.

Integrated Gradients. The resulting maps display varying levels of spatial focus. While some foreground regions show elevated attribution, there is no consistent correspondence to distinct anatomical features associated with lung pathologies. These results reflect the methods sensitivity to input structure and the absence of explicit spatial supervision during model training.

Grad-CAM. We applied Grad-CAM to the final layers of the spatial and causal transformers, as well as to the VQ layer of the CT-CLIP encoder. Attribution maps from the spatial and causal layers were typically sparse and diffuse, with activations often appearing in anatomically irrelevant regions, including image padding. These patterns lacked consistent correspondence with pathology-related structures, indicating limited utility for weak localization.

Grad-CAM applied to the VQ layer yielded more spatially contiguous activations, likely due to its role in aggregating and discretizing global features. However, these maps frequently emphasized background or non-lung regions, suggesting the models reliance on scanner-specific artifacts or formatting cues rather than clinically meaningful content. This behavior indicates potential overfitting to dataset-specific structure and undermines the generalizability of Grad-CAM-based interpretability in this setting.

Occlusion Sensitivity. The resulting maps exhibit more localized and interpretable patterns, with activations predominantly focused on the lung regions. The signal is typically weak or absent across most of the scan, but becomes distinctly stronger within slices containing lung tissue, before diminishing again once the lungs are no longer visible. While occasional signals appear outside the lungs, they rarely occur in background regions, making this approach the most consistent and reliable among the attribution methods evaluated.

5 Discussion

The strongest results were obtained using the occlusion method—the only model-agnostic attribution approach in our study. This method consistently highlighted localized, high-signal regions within the lungs while largely ignoring irrelevant scan areas. These findings suggest that VLP models may have potential for localizing pathologies described in accompanying text, even within volumetric data. However, to rigorously evaluate the quality of the resulting attribution maps, a follow-up study should be conducted using a dataset with ground truth segmentation labels. This would require either annotating portions of the CT-RATE dataset or constructing a new dataset with segmentation masks and training a CLIP-style model on it.

In contrast, model-specific attribution methods were largely ineffective in producing meaningful spatial explanations. At best, gradient-based techniques such as Integrated Gradients and Grad-CAM highlighted broad, non-specific regions of the body without clearly localizing pathology-relevant areas. At worst, the attribution maps appeared scattered or random, lacking any apparent anatomical correspondence. These findings highlight the uncertainty around whether the gradient signal derived from cosine similarity between image and text embeddings is sufficiently informative to enable spatial localization in complex 3D medical images.

Attention-based methods, which do not rely on gradients, also failed to produce spatially meaningful attribution maps. In our experiments, attention rollout resulted in extremely sparse or diffuse activations across the volume, and individual attention heads often produced patterns resembling noise rather than localized structure. These findings are consistent with those of Zhao et al. [18], who evaluated a range of attribution techniques on 2D CLIP models in the natural image domain. They attributed the poor performance of attention-based methods to the inherent sparsity of attention distributions in transformers.

We theorize that the poor quality of attribution maps may stem, in part, from the architectural structure of CT-CLIPs image encoder. Rather than allowing attention to flow freely across the entire scan, the model applies a two-stage transformer: a spatial transformer, which restricts attention within individual axial slices, followed by a causal transformer, which aggregates information across depth-wise pillars of patches. While this setup enables each patch to eventually access global context, the separation into two rigid stages introduces artificial constraints that may distort spatial relationships—ultimately degrading the interpretability of attention-based visualizations.

This architectural choice can be traced back to the origins of CT-ViT, which was adapted from a video model architecture [13]. In the video domain, separating spatial and causal transformers makes intuitive sense: one processes individual frames, while the other captures motion dynamics. Although CT volumes are three-dimensional, treating them as temporally structured sequences may not be ideal. The two-part transformer design was likely motivated by the need to reduce the quadratic cost of self-attention. However, alternative architectures—such as a single 3D Swin Transformer—could offer a more holistic representation of the scan, while maintaining tractable attention computation through localized attention windows [19].

6 Conclusion

In this work, we explored the potential of CLIP-style vision–language models for weakly supervised localization in 3D chest CT scans using free-text radiology reports. By adapting several attribution methods, we assessed whether the learned image–text alignment encodes spatially meaningful features. While the occlusion method produced the most promising and localized attribution maps, gradient- and attention-based techniques struggled to reveal coherent patterns—likely due to architectural constraints and the global nature of contrastive supervision. Our results suggest that vision–language pretraining can support localization in volumetric medical data, but further progress may require architectural adjustments.

Acknowledgements. This project has received funding from the European Innovation Council (EIC) under the European Union's Horizon Europe research and innovation programme (grant agreement No 190191356). This work has also been funded by the European Union and Ministry of Education and Research via project TEM-TA72.

Disclosure of Interest. The authors have no competing interests to declare that are relevant to the content of this article.

References

1. Sistaninejhad, B., Rasi, H., Nayeri, P.: A review paper about deep learning for medical image analysis. Comput. Math. Methods Med. **2023**, 7091301 (2023)
2. Litjens, G., et al.: A survey on deep learning in medical image analysis. Med. Image Anal. **42**, 60–88 (2017)
3. Hamamci, I.E., et al.: Developing generalist foundation models from a multimodal dataset for 3d computed tomography. arXiv preprint arXiv:2403.17834 (2024)
4. Chen, Y., Liu, C., Liu, X., Arcucci, R., Xiong, Z.: BIMCV-R: a landmark dataset for 3D CT text-image retrieval. In: Linguraru, M.G., et al. (eds.) Medical Image Computing and Computer Assisted Intervention – MICCAI 2024. MICCAI 2024. LNCS, vol. 15011. Springer, Cham . (2024).https://doi.org/10.1007/978-3-031-72120-5_12
5. Radford, A., et al. Learning transferable visual models from natural language supervision. arXiv preprint arXiv: 2103.00020 (2021)
6. Zhao, Z., et al.: Clip in medical imaging: a survey. Med. Image Anal., 103551 (2025)
7. Lei, Y., Li, Z., Shen, Y., Zhang, J., Shan, H.: Clip-lung: textual knowledge-guided lung nodule malignancy prediction. In: International Conference on Medical Image Computing and Computer-Assisted Intervention, pp. 403–412. Springer (2023). https://doi.org/10.1007/978-3-031-43990-2_38
8. Lin, J., et al.: Ct-glip: 3d grounded language-image pretraining with ct scans and radiology reports for full-body scenarios. arXiv preprint arXiv:2404.15272 (2024)
9. Abnar, S., Zuidema, W.: Quantifying attention flow in transformers. arXiv preprint arXiv: 2005.00928 (2020)
10. Lin, Y., et al.: Clip is also an efficient segmenter: a text-driven approach for weakly supervised semantic segmentation. In: 2023 IEEE/CVF Conference on Computer Vision and Pattern Recognition (CVPR), pp. 15305–15314 (2023)
11. Wang, H., et al.: Score-cam: score-weighted visual explanations for convolutional neural networks. In: Proceedings of the IEEE/CVF Conference on Computer Vision and Pattern Recognition Workshops, pages 24–25 (2020)
12. Chen, P., Li, Q., Biaz, S., Bui, T., Nguyen, A.: gscorecam: what objects is clip looking at? In: Proceedings of the Asian Conference on Computer Vision, pp. 1959–1975 (2022)
13. Hamamci, I.E. et al.: GenerateCT: text-conditional generation of 3D chest CT volumes. In: Leonardis, A., Ricci, E., Roth, S., Russakovsky, O., Sattler, T., Varol, G. (eds.) Computer Vision – ECCV 2024. ECCV 2024. LNCS, vol. 15137. Springer, Cham (2025).https://doi.org/10.1007/978-3-031-72986-7_8
14. Boecking, B. et al.: Making the most of text semantics to improve biomedical vision–language processing. In: Avidan, S., Brostow, G., Cissé, M., Farinella, G.M., Hassner, T. (eds.) Computer Vision – ECCV 2022. ECCV 2022. LNCS, vol. 13696. Springer, Cham (2022). https://doi.org/10.1007/978-3-031-20059-5_1

15. Sundararajan, M., Taly, A., Yan, Q.: Axiomatic attribution for deep networks. In: International Conference on Machine Learning, pp. 3319–3328. PMLR (2017)
16. Selvaraju, R.R., Cogswell, M., Das, A., Vedantam, R., Parikh, D., Batra, D.: Gradcam: visual explanations from deep networks via gradient-based localization. In: Proceedings of the IEEE international Conference on Computer Vision, pp. 618–626 (2017)
17. Zeiler, M.D., Fergus, R.: Visualizing and understanding convolutional networks. In: Fleet, D., Pajdla, T., Schiele, B., Tuytelaars, T. (eds.) ECCV 2014. LNCS, vol. 8689, pp. 818–833. Springer, Cham (2014). https://doi.org/10.1007/978-3-319-10590-1_53
18. Zhao, C., Wang, K., Hsiao, J.H., Chan, A.B.: Grad-eclip: Gradient-based visual and textual explanations for clip. arXiv preprint arXiv:2502.18816, (2025)
19. atamizadeh, A., Nath, V., Tang, Y., Yang, D., Roth, H.R., Xu, D.: Swin UNETR: swin transformers for semantic segmentation of brain tumors in MRI images. In: Crimi, A., Bakas, S. (eds.) Brainlesion: Glioma, Multiple Sclerosis, Stroke and Traumatic Brain Injuries. BrainLes 2021. LNCS, vol. 12962. Springer, Cham (2022). https://doi.org/10.1007/978-3-031-08999-2_22

Automated Constraint-Aware X-ray View Planning for Vascular Interventions Using Preoperative CTA

Baochang Zhang[1,2,3]($\boxtimes$), Abdelkader Saad[1], Heribert Schunkert[2,4], and Nassir Navab[1,3]

[1] Computer Aided Medical Procedures, Technical University of Munich, Munich, Germany
baochang.zhang@tum.de
[2] German Heart Center Munich, Munich, Germany
[3] Munich Center for Machine Learning, Munich, Germany
[4] German Centre for Cardiovascular Research, Munich Heart Alliance, Munich, Germany

Abstract. Accurate intraoperative imaging is essential for successful endovascular aneurysm repair (EVAR), enabling navigation of complex vascular anatomies and precise device placement. Surgeons often acquire multiple angiographic views, but manual viewpoint selection can lead to repeated C-arm repositioning, increased radiation exposure, and prolonged procedures. While recent methods automate view planning using vascular geometry and pose estimation, they often assume unrestricted C-arm mobility and overlook device-specific spatial constraints. In this work, we propose a novel constraint-aware, automated multi-view planning framework that leverages preoperative CTA data to generate optimized X-ray views tailored to procedural and equipment limitations. Our method starts with vessel segmentation, centerline extraction, and vessel graph construction. A planning route is defined along the target centerline, from which discrete points are sampled as local region centers. For each center, we define a region of interest and solve a constrained optimization problem to determine the optimal viewing orientation. The objective function combines two criteria: vessel spread area, computed via the convex hull area of the projected centerline, and inter-region projection separation, which promotes spatially clear views by minimizing overlap. We validated our framework on an in-house preoperative CTA dataset from 27 patients. Both qualitative and quantitative results demonstrate improved region visibility, spatial separation, and continuity of optimal viewing poses along the vascular path.

Keywords: View Planning · Constraint-Aware Optimization · Endovascular Procedures

---

B. Zhang and A. Saad—The two authors contributed equally to this paper.

1 Introduction

Minimally invasive vascular interventions such as endovascular aneurysm repair (EVAR) rely heavily on high-quality intraoperative imaging for successful navigation and device placement [3,10]. In these procedures, clear visualization of the vascular anatomy, particularly around bifurcations, curvatures, and aneurysmal segments, is critical. Surgeons typically acquire multiple angiography X-ray views to guide catheters through complex vascular structures. However, manual selection of these views remains subjective and inconsistent, which leads to excessive C-arm repositioning, prolonged procedure times, and increased exposure to radiation and contrast agents [2,5,11].

Recent advances in medical image analysis and multi-modality integration have paved the way for more intelligent and efficient imaging workflows. Preoperative CT angiography (CTA) provides detailed anatomical information, enabling the development of machine learning or deep learning methods for tasks such as vessel segmentation [4] and bifurcation detection [8]. For intraoperative X-ray view estimation, several approaches based on geometric heuristics or C-arm positioning feedback have been proposed [2,7,13], often involving plane fitting to centerline segments or cross-modality view matching. While these methods show promise, many assume unconstrained C-arm motion, and rely on curated training data or manual initialization [5,11]. In practice, spatial constraints such as limited degrees of freedom, bulky gantries, and sterile field requirements complicate view planning, limiting the applicability of these idealized approaches.

To address these limitations, we propose a constraint-aware optimization framework for automated X-ray view planning in vascular interventions. Our method uses geometric priors extracted from preoperative CTA, specifically vessel centerlines, to define region-specific objectives that balance two key criteria: maximizing the projected coverage of vessel anatomy within each local region and promoting clear spatial separation between overlapping vascular segments in the projection plane. Local anatomical regions are weighted based on predefined clinical priorities, and a customized scoring function guides the optimization to emphasize critical structures. The optimization problem is formulated as a constrained maximization of this composite score over bounded orientation intervals, reflecting the mechanical constraints and limited degrees of freedom of clinical C-arm systems. Non-differentiability arises from the discrete sampling of vessel regions and non-smooth scoring components. To solve this non-differentiable problem efficiently, we employ the BOBYQA algorithm [6], a derivative-free method well-suited for applications where gradient information is unavailable or unreliable. This formulation ensures clinically feasible and smooth pose transitions while avoiding the need for learned models or RANSAC-based estimators.

Our key contributions are: (i) a mathematically rigorous formulation for constraint-aware X-ray view planning that does not depend on annotated training data or predefined local planes; (ii) a novel region-based scoring function integrating clinical priorities with geometric constraints; and (iii) a comprehensive validation demonstrating improved vascular visibility, enhanced spatial separation of vessel regions, and coherent pose trajectories along the intervention

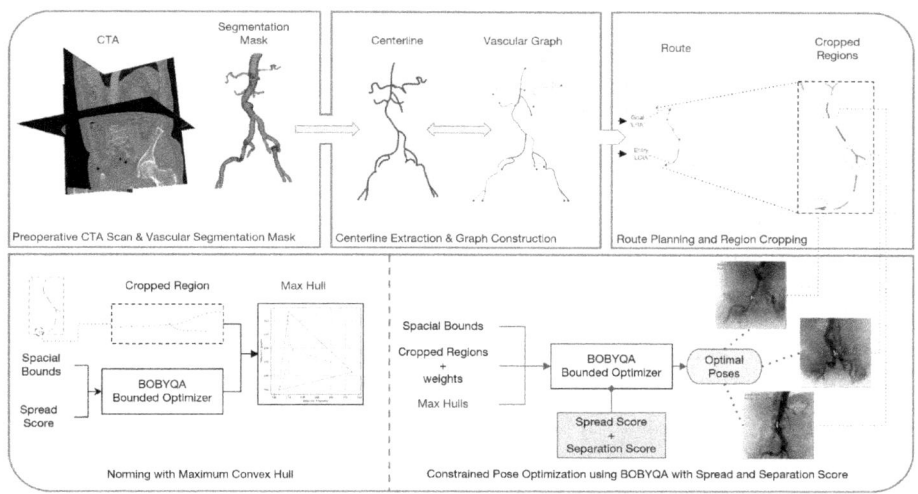

Fig. 1. An overview of the proposed pipeline for automatic constraint-aware view planning for vascular interventions.

path. These results confirm the potential of our approach to enhance procedural efficiency while respecting clinical constraints.

2 Methodology

This section provides a detailed description of the methodology outlined in Fig. 1.

2.1 Topology Information Extraction

Vascular Graph Construction. We first segment the vascular structures from the preoperative CTA scan using nnU-Net [4,13], followed by centerline extraction to capture the vessel topology. These centerlines are used to construct an undirected graph $G = (V, E)$, where each node $v \in V$ represents an anatomical landmark, such as a bifurcation, an endpoint or an aneurysm's centroid, and each edge $e \in E$ corresponds to a vessel segment connecting two landmarks. In this graph, the degree of a node is defined as the number of edges connected to it. Endpoints are nodes with a degree of one, indicating a connection to only one other node, while bifurcations are nodes with a degree greater than two, reflecting a branching point in the vessel network. Following the method described in [13], the aneurysms can be segmented via nnU-Net [4], and their centroids are computed and added to the graph as annotated nodes. This representation preserves both geometric structure and anatomical context for downstream processing.

Route Planning. Given the vascular graph $G = (V, E)$, a starting node v_s and goal node v_t, Dijkstra's algorithm [1] is applied to compute the shortest intervention route $P = (V', E')$. The resulting route consists of a sequence of anatomical landmarks $V' = \{v_s, v_1, v_2, \ldots, v_t\}$, which are then traversed in order along the intervention route.

Reference Points Sampling. In order to capture the curvature of vessel branches between these landmarks V', the reference points for the downstream view optimization should include not only these landmarks but also intermediate points sampled along the vessel centerlines E'. Therefore, a fixed spacing parameter d_s (in millimeters) defines the uniform sampling density. Each centerline segment between consecutive nodes in the planned route is parameterized by arc length as:

$$\ell_i = \sum_{j=1}^{i} \|\mathbf{x}_j - \mathbf{x}_{j-1}\|_2, \quad \text{for } i = 1, \ldots, N \tag{1}$$

where $\mathbf{x}_j$ are the centerline points and ℓ_i is the arc-length at point i. Linear interpolation over this arc-length parameterization is then used to sample new points at uniform intervals of d_s. Sampling is performed independently for each segment to ensure consistent and anatomically meaningful coverage along the intervention route.

Local Topology Extraction. After defining the reference points along the intervention route, we extract local vascular regions around each point to guide the X-ray view optimization. These regions are selected based on both the vessel topology and spatial proximity. For points located on vessel branches, the local region includes only the corresponding vessel segment. For points at bifurcations, the region is expanded to include all directly connected neighboring branches, ensuring the full branching structure is captured and reducing overlap at bifurcations in projection. To avoid the influence of long distant vessel structures with complex curvature, we limit each local region to a fixed 3D radius r_c centered at the reference point. Additionally, to prevent overlap between adjacent regions, this radius is constrained to be no larger than half the sampling distance d_s.

2.2 View Optimization

With the local regions of interest defined, we now optimize the X-ray viewing pose to enhance vascular visibility. In unconstrained settings, a common approach is to compute the mean plane [13] of the target vessel structure and align the C-arm such that the X-ray beam is orthogonal to this plane. However, in real-world operating rooms, mechanical limits often restrict the range of motion of the imaging system. As such, these idealized views may be physically unreachable. To overcome this, we introduce a constraint-aware optimization strategy that prioritizes broad vascular coverage and clear separation of relevant structures in the resulting projection. Here, we represent each C-arm view as a pose in 3D space with six degrees of freedom:

$$\theta = (\boldsymbol{\omega}, \mathbf{t}) = (\alpha, \beta, \gamma, t_x, t_y, t_z) \qquad (2)$$

where $\boldsymbol{\omega} = (\alpha, \beta, \gamma)$ is the orientation (Euler angles), and $\mathbf{t} = (t_x, t_y, t_z)$ is the translation. Since translation of the C-arm is typically unconstrained in clinical settings, we set the translation vector $\mathbf{t}$ to the 3D coordinates of the center of the current local region to maintain focus on the anatomical region of interest. Then, the optimization is performed over the orientation vector $\boldsymbol{\omega}$, constrained to a bounded domain Ω, reflecting the mechanical limits of the imaging system:

$$\boldsymbol{\omega} \in \Omega = [\alpha_{\min}, \alpha_{\max}] \times [\beta_{\min}, \beta_{\max}] \times [\gamma_{\min}, \gamma_{\max}] \qquad (3)$$

Objective Function. In our proposed constraint-aware optimization strategy, the objective function integrates two complementary criteria: (i) the spatial coverage of the vessel region on the 2d detector plane, quantified by S_{spread}; and (ii) the separation between the projected images of different anatomical regions, measured by S_{sep}.

To compute the spread score S_{spread}, each local region is projected onto the detector plane. The area of the convex hull enclosing the resulting 2D points is then used to quantify how widely the region is distributed in the image. The convex hull $conh(S)$ of a point set $S = \{x_1, x_2, \ldots, x_n\}$ is defined as the smallest convex shape that contains all the points in S, and is formally given by:

$$conh(S) = \left\{ \sum_{i=1}^{n} \lambda_i x_i \;\middle|\; \lambda_i \geq 0, \sum_{i=1}^{n} \lambda_i = 1 \right\}. \qquad (4)$$

The set of points S in this context corresponds to the projection of a region R onto the 2D detector plane under the pose θ. Hence, the convex hull surface area $A_{\text{hull},i}(\theta)$ serves as a measure of how much the region R_i spreads in projection under pose θ

$$A_{\text{hull},i}(\theta) = \text{Area}\left(conh(\pi_\theta(R_i))\right). \qquad (5)$$

Since the convex hull area also depends on the underlying shape and topology of each 3D local region, we normalize the area of each region's convex hull to ensure that all regions are treated equally during optimization. Specifically, we normalize the area using the maximal achievable convex hull area under spatial constraints, defined as $A_{\text{hull},i}^{\max} = \max_{\theta} A_{\text{hull},i}(\theta)$. Meanwhile, each region can be assigned a clinical priority factor p_i, where lower priority values p_i indicate higher importance. Finally, the spread score can be defined as:

$$S_{\text{spread}}(R, p, \kappa, \theta) = \sum_{i=1}^{N} \lambda_i \cdot \frac{A_{\text{hull},i}(\theta)}{A_{\text{hull},i}^{\max}}, \quad \text{with} \quad \lambda_i = (1 + p_{\max} - p_i)^\kappa \qquad (6)$$

where $\kappa = 2$ modulates its contribution to the spread score, $p_{\max}$ means the maximum value among the priority set p, and N represents the number of regions participating in view optimization.

The separation score S_{sep} is defined as the minimum pairwise distance between the 2D centroids of the projected regions. This score encourages view configurations in which the regions appear more spatially distinct on the detector plane.

$$S_{\text{sep}}(R, \theta) = \min_{i \neq j} \|\mathbf{u}_i - \mathbf{u}_j\|_2 \tag{7}$$

where $\mathbf{u}_i$ denotes the centroid of the projected 2D points of region R_i. By combining the two scores through a weighted sum, we define the total objective function as:

$$S_{\text{total}}(R, p, \kappa, \theta) = \nu_s \cdot S_{\text{spread}}(R, p, \kappa, \theta) + \nu_{sc} \cdot S_{\text{sep}}(R, \theta), \tag{8}$$

where ν_s and ν_{sc} are weighting coefficients that balance the contributions of the spread and separation scores, respectively.

Constrained Optimization. Given the objective function S_{total} and the bounded domain Ω, we solve

$$\boldsymbol{\omega}^* = \arg\max_{\boldsymbol{\omega} \in \Omega} S_{\text{total}}(R, p, \kappa, (\boldsymbol{\omega}, \mathbf{t})) \tag{9}$$

to determine the optimal X-ray viewing angles along the vascular route. This formulation involves components such as convex hull area and minimum inter-region distances, which are non-differentiable. To handle this, we employ the derivative-free BOBYQA optimizer [6], which is well-suited for optimizing non-differentiable functions within bounded domains. During traversal of the vascular route, each optimization is initialized using the optimal orientation from the previous region. This promotes temporal continuity and reduces abrupt transitions between successive views. The final optimal pose is defined by the optimized orientation in combination with the fixed translation, expressed as:

$$\theta^* = (\boldsymbol{\omega}^*, \mathbf{t}), \tag{10}$$

which is subsequently used to generate a digital reconstructed radiography (DRR) image [9,12] for visualization.

3 Experiments and Results

3.1 Experimental Setup and Dataset

Reference points are sampled along the vessel centerlines at fixed intervals of $d_s = 40$ mm. For local region cropping, the fixed 3D radius r_c is set as 26 mm. For DRR generation, a virtual X-ray imaging system is used, configured to match the geometric parameters of a standard C-arm device. Specifically, the distance from the X-ray source to the isocenter is 742.5 mm, and the distance from the detector to the isocenter is 517.15 mm. The detector has a width and height of 432 mm, with a pixel size of 0.3 mm.

A abdominal dataset collected from 27 patients diagnosed with aneurysms is used for experiments. For each patient, a preoperative CTA scan was acquired using a GE Revolution EVO CT scanner. The reconstructed slice thickness of the CTA images ranges from 1 mm to 3 mm, and the in-plane spacing ranges from 0.79 mm to 1.34 mm.

3.2 Visual Results

We compare our method against two commonly used view planning strategies. The first is the mean-plane view, where the C-arm is oriented orthogonally to the mean plane of the target vessel segment, following the approach of [13]. The second is a clinical base view, a static configuration using default pose angles that simulates standard clinical practice without any optimization. To reflect the realistic mechanical limits of the C-arm system, we impose angular constraints of $\pm 20°$, $\pm 20°$, and $\pm 10°$ on the rotation angles α, β, and γ, respectively.

The comparison was performed using DRRs generated under the poses prescribed by each method. As shown in Fig. 2, our method produces DRRs with improved anatomical coverage and clearer separation of vascular structures, particularly around bifurcations and curved regions. The green overlay highlights the primary target region in each view, with the optimized pose parameters displayed in the top-left corner of the image produced by our method. Our method generates projections that resemble clinical views in many cases but reveal additional curvature or elongation in areas that are otherwise flattened. Notably, in some regions where the orthogonal views method produces poses that exceed the spatial constraints, our optimized views maintain vessel spread within the prescribed mechanical limits.

3.3 Impact of Initialization and Prioritization on Displacement

We evaluated the stability and continuity of C-arm pose transitions along a vascular route consisting of eight regions, beginning from the anteroposterior (AP) view. Displacement between sequential poses was measured as the element-wise absolute angular difference between successive optimal orientations, defined as $\sum_i |\omega_i^* - \omega_{i-1}^*|$ across the three rotation axes α, β, and γ.

We compared multiple view optimization strategies using either 0-NN or 1-NN, under two initialization schemes, as shown in Table 1. In init-1, each optimization step is warm-started using the previously computed optimal pose. In init-2, optimization is reinitialized from the AP view at each step. "K-NN" refers to using the K Nearest Neighboring regions along the route during each optimization step. All experiments were conducted under relaxed angular bounds of $\pm 45°$, allowing broader exploration of pose variability.

Among all configurations, the 1-NN&init1 strategy yielded the most stable transitions, with the lowest average displacements per axis (4.3°, 7.2°, and

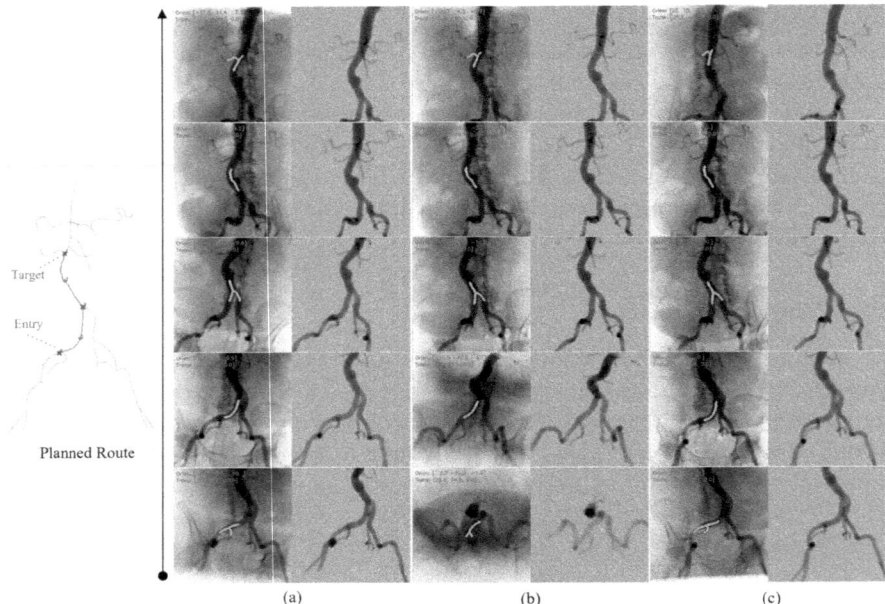

Fig. 2. Comparison of DRRs generated using (a) our view optimization method, (b) the orthogonal views method [13] and (c) clinical base view without any optimization.

Table 1. Per-Axis Displacement for Optimization Strategies

Metric in deg (α, β, γ)	1-NN&init1			1-NN&init2			0-NN&init1			0-NN&init2		
Total displ. per axis	30.1	50.6	144.2	180.6	124.4	86.3	219.9	207.4	66.5	224.9	187.9	128.7
Avg displ. per axis	4.3	7.2	20.6	25.8	17.8	12.3	31.4	29.6	9.5	32.1	26.8	18.4
Std Dev	3.5	12.1	25.3	30.4	11.5	22.5	23.4	24.1	16.6	26.4	15.8	16.0

20.6°). This suggests that incorporating local continuity along the route and warm-starting from the previous pose significantly reduces inter-view variability. In contrast, the 0-NN&init2 strategy led to the highest displacement (average: 32.1°, 26.8°, 18.4°), indicating that ignoring neighboring regions and reinitializing from a fixed AP view can produce abrupt pose shifts. Overall, these results underscore the importance of continuity-aware initialization and spatially informed multi-region optimization.

4 Conclusion

In this work, we proposed a constraint-aware optimization framework for C-arm view planning that simultaneously maximizes vascular region coverage and spatial separation under realistic mechanical limits. By integrating convex hull-based spread and inter-region separation scores into a unified objective, our method effectively identifies optimal X-ray poses that improve anatomical visibility while respecting device constraints. The use of derivative-free optimization and warm-starting initialization ensures smooth transitions along vascular paths, enhancing clinical applicability. Our experiments on abdominal vascular datasets demonstrated that the optimized views consistently outperform common baseline strategies, such as orthogonal mean-plane and static clinical base views, by providing clearer separation and better spatial coverage of vascular structures. Overall, our approach offers a practical solution to improve intraoperative imaging and navigation by providing optimized, constraint-compliant X-ray viewpoints. Future work will explore adaptive weighting schemes and real-time implementation to further support dynamic clinical scenarios.

Acknowledgments. The project was supported by the Bavarian State Ministry of Science and Arts within the framework of the "Digitaler Herz-OP" project under the grant number 1530/891 02 and the China Scholarship Council (File No.202004910390). We also thank BrainLab AG for their partial support.

Disclosure of Interests. The authors have no competing interests to declare that are relevant to the content of this article.

References

1. Dijkstra, E.W.: A note on two problems in connexion with graphs. In: Edsger Wybe Dijkstra: His Life, Work, And Legacy, pp. 287–290 (2022)
2. Fallavollita, P., et al.: *Desired-View* controlled positioning of angiographic c-arms. In: Golland, P., Hata, N., Barillot, C., Hornegger, J., Howe, R. (eds.) MICCAI 2014. LNCS, vol. 8674, pp. 659–666. Springer, Cham (2014). https://doi.org/10.1007/978-3-319-10470-6_82
3. Hinchliffe, R., Ivancev, K.: Endovascular aneurysm repair: current and future status. Cardiovasc. Intervent. Radiol. **31**, 451–459 (2008)
4. Isensee, F., Jaeger, P.F., Kohl, S.A., Petersen, J., Maier-Hein, K.H.: nnu-net: a self-configuring method for deep learning-based biomedical image segmentation. Nat. Methods **18**(2), 203–211 (2021)
5. Kausch, L., et al.: Toward automatic c-arm positioning for standard projections in orthopedic surgery. Int. J. Comput. Assist. Radiol. Surg. **15**, 1095–1105 (2020)
6. Powell, M.J., et al.: The bobyqa algorithm for bound constrained optimization without derivatives. Cambridge NA Report NA2009/06, University of Cambridge, Cambridge **26**, 26–46 (2009)
7. Tehlan, K., Winkler, A., Roth, D., Navab, N.: X-ray device positioning with augmented reality visual feedback. In: 2022 IEEE Conference on Virtual Reality and 3D User Interfaces Abstracts and Workshops (VRW), pp. 870–871. IEEE (2022)
8. Tetteh, G., et al.: Deepvesselnet: vessel segmentation, centerline prediction, and bifurcation detection in 3-d angiographic volumes. Front. Neurosci. **14**, 592352 (2020)

9. Unberath, M., et al.: DeepDRR – a catalyst for machine learning in fluoroscopy-guided procedures. In: Frangi, A.F., Schnabel, J.A., Davatzikos, C., Alberola-López, C., Fichtinger, G. (eds.) MICCAI 2018. LNCS, vol. 11073, pp. 98–106. Springer, Cham (2018). https://doi.org/10.1007/978-3-030-00937-3_12
10. Veulemans, V., et al.: Optimal c-arm angulation during transcatheter aortic valve replacement: accuracy of a rotational c-arm computed tomography based three dimensional heart model. World J. Cardiol. **8**(10), 606 (2016)
11. Virga, S., Dogeanu, V., Fallavollita, P., Ghotbi, R., Navab, N., Demirci, S.: Optimal C-arm positioning for aortic interventions. In: Handels, H., Deserno, T.M., Meinzer, H.-P., Tolxdorff, T. (eds.) Bildverarbeitung für die Medizin 2015. I, pp. 53–58. Springer, Heidelberg (2015). https://doi.org/10.1007/978-3-662-46224-9_11
12. Zhang, B., et al.: A patient-specific self-supervised model for automatic x-ray/ct registration. In: International Conference on Medical Image Computing and Computer-Assisted Intervention, pp. 515–524. Springer (2023). https://doi.org/10.1007/978-3-031-43996-4_49
13. Zhang, B., Liu, Y., Liu, S., Schunkert, H., Ghotbi, R., Navab, N.: Automated multi-view planning for endovascular aneurysm repair procedures. In: Clinical Image-Based Procedures (CLIP 2024), LNCS, vol. 15196, pp. 22–31. Springer (2024). https://doi.org/10.1007/978-3-031-73083-2_3

Author Index

A
Al Turkestani, Najla 42
Aliaga, Aron 42
Ariva, Joonas 87

B
Bale, Reto 65
Barone, Selene 42
Bianchi, Jonas 42

C
Caleme, Eduardo 42
Cevidanes, Lucia 42
Chelebian, Eduard 21
Chen, Yufei 11, 53
Choi, Kyu Sung 32

D
Dai, Runpeng 42
DaSilva, Alexandre F. 42

E
Egebjerg, Kristian 21

F
Fan, Panpan 11
Fishman, Dmytro 87
Fogarollo, Stefano 65
Freysinger, Wolfgang 65

G
Gaydamour, Alban 42
Goncalves, Daniela 42
Goncalves, Joao 42
Gurgel, Marcela 42

H
Hanauer, David 42
Hershey, Adam 42
Hsu, Nina 42

I
Injarabian, David Avedis 87

J
Jang, Han 32
Joshi, Devvrat 75

K
Kwon, Anabelle 42

L
Lee, Junhyeok 32
Li, Tengfei 42
Liu, Wei 53
Lukes, Adela 65

M
Martin, J. Ryan 1
Mathian, Émilie 21
Mattos, Claudia 42
Miranda, Felicia 42
Moyer, Daniel 1
Mund, Andreas 21

N
Navab, Nassir 97

O
Oldenburg, Lukas 21

P
Pieper, Steve 42
Prieto, Juan 42

R
Rekik, Islem 75
Ruellas, Antonio 42

S
Saad, Abdelkader 97
Schunkert, Heribert 97
Schweizer, Lisa 21
Shen, Qiyun 53
Shi, Jie 11
Shin, Maxwell 42
Strauss, Maximillian T. 21
Styner, Martin 42
Suh, Yehyun 1
Šuvalov, Hendrik 87

T
Teixeira, Rodrigo 42
Tulissi, Enzo 42

U
Ummat, Ishani 21

W
Wei, Zhipeng 11
Wolford, Lawrence 42

X
Xu, Zhikang 11

Y
Yatabe, Marilia 42
Yue, Xiaodong 11

Z
Zhang, Baochang 97
Zhang, Qi 53
Zhou, Kefan 53
Zhu, Hongtu 42
Zonderland, Gijs 21
Zupelari, Marina 42
Zupelari, Paulo 42

GPSR Compliance

The European Union's (EU) General Product Safety Regulation (GPSR) is a set of rules that requires consumer products to be safe and our obligations to ensure this.

If you have any concerns about our products, you can contact us on

ProductSafety@springernature.com

In case Publisher is established outside the EU, the EU authorized representative is:

Springer Nature Customer Service Center GmbH
Europaplatz 3
69115 Heidelberg, Germany

www.ingramcontent.com/pod-product-compliance
Lightning Source LLC
Chambersburg PA
CBHW070445071025
33641CB00035B/893